VISUALISE

VISUALISE

THINK

FEEL

PERFORM

LIKE THE TOP 1%

MAYA RAICHOORA

RIDER

RIDER

UK | USA | Canada | Ireland | Australia
India | New Zealand | South Africa

Rider is part of the Penguin Random House group of companies
whose addresses can be found at global.penguinrandomhouse.com

Penguin Random House UK
One Embassy Gardens, 8 Viaduct Gardens, London SW11 7BW

penguin.co.uk
global.penguinrandomhouse.com

First published by Rider in 2025

1

Copyright © Maya Raichoora 2025
Illustration © Hannah Williams
The moral right of the author has been asserted.

No part of this book may be used or reproduced in any manner for the purpose
of training artificial intelligence technologies or systems. In accordance with
Article 4(3) of the DSM Directive 2019/790, Penguin Random House
expressly reserves this work from the text and data mining exception.

The information in this book has been compiled as general guidance on visualisation
techniques. It is not a substitute and not to be relied on for medical or healthcare
professional advice. Please consult your medical practitioner before changing,
stopping or starting any medical treatment. So far as the author is aware the
information given is correct and up to date as at 27 January 2025. The author
and publishers disclaim, as far as the law allows, any liability arising directly or
indirectly from the use or misuse of the information contained in this book.

The names of people and the detail of events have been
changed to protect the privacy of others.

Typeset in 10.75/16pt ITC Galliard Pro by Jouve (UK), Milton Keynes
Printed and bound in India by Thomson Press India Ltd.

The authorised representative in the EEA is Penguin Random House Ireland,
Morrison Chambers, 32 Nassau Street, Dublin D02 YH68

A CIP catalogue record for this book is available from the British Library

ISBN 9781846048524

Penguin Random House is committed to a sustainable future
for our business, our readers and our planet. This book is made
from Forest Stewardship Council® certified paper.

*To my 15-year-old self, for dreaming beyond the limits.
And to my future 50-year-old self, for proving that imagination is a force that never grows old.*

CONTENTS

Introduction   1
My Story   7
How to Use This Book   16

PART 1
AWARENESS: LEARNING THE FOUNDATIONS

CHAPTER 1 – HOW VISUALISATION CAN REWIRE YOUR BRAIN   21
Neuroscience 101   23
The Brain Is Like a City   30

CHAPTER 2 – PAY ATTENTION TO YOUR THOUGHTS   33
The Myth of 'Always Think Positive'   35
The Power of the Human Imagination   38

CHAPTER 3 – FEEL IT TO HEAL IT   42
Emotional Awareness   44
Emotional Intelligence   45
Emotional Agility   46
Expressing Your Emotions   48

CHAPTER 4 – OVERCOME LIMITING BELIEFS	50
Cultivating Self-Belief	53
The Number-One Reason You Aren't Achieving Your Goals	59

CHAPTER 5 – THE POWER OF VISUALISATION	62
Visualisation Versus Meditation and Mindfulness	66
Visualisation Versus Manifestation	67

CHAPTER 6 – BUILDING A CHAMPION'S MIND	72
What I Have Learned from Elite Athletes	73
Developing Unshakeable Confidence	86

CHAPTER 7 – SETTING AND ACHIEVING GOALS	93
What Do I Want?	94
Why Do I Want It?	97
Who Do I Need to Be to Achieve It?	98
Live Like a Predator, Not Like Prey	99

PART 2
INTELLIGENCE: MASTERING THE FIVE VISUALISATION TECHNIQUES

CHAPTER 8 – OUTCOME VISUALISATION	107
Honing Your Goals and Vision	108
Increasing Self-Belief	113
Tapping Into Unstoppable Confidence	115
Generating Hope and Resilience	121
Increasing Motivation	124

Overcoming Doubt and Fear	126
What to Expect	127

CHAPTER 9 – PROCESS VISUALISATION — 130

Performing Like the Top 1 Per Cent	131
Overcoming Performance Anxiety	136
Increasing Focus and Productivity	141
Mentally Rehearsing Your Day	144
Enhancing Sports and Fitness	146
Breaking and Building Habits	149
Letting Go	152
What to Expect	154

CHAPTER 10 – CREATIVE VISUALISATION — 156

Managing Difficult Emotions	157
Feeling Gratitude and Joy	161
Connecting to Your Inner Confidence	162
Managing Pain	163
Relieving Stress	165
What to Expect	168

CHAPTER 11 – NEGATIVE VISUALISATION — 171

Enhancing Performance and Being Prepared	172
Encouraging Appreciation and Motivation	175
Achieving Goals	175
What to Expect	176

CHAPTER 12 – EXPLORATIVE VISUALISATION — 178

Creating Content	180
Designing for Speeches, Events and Presentations	183
Making Decisions and Solving Problems	184

Generating New Ideas and Creativity   186
What to Expect   189

PART 3
EXECUTION: PURSUING YOUR EXCELLENCE

CHAPTER 13 – BUILDING YOUR ROUTINE   193
When   193
Where   196
How Long For   197
Tailoring Your Visualisations   198
Building a Visualisation Habit   199
You Reap What You Repeat   202
The Painkiller Versus Supplement Mentality   203
Add Some Toppings   204

CHAPTER 14 – FALL IN LOVE WITH DISCIPLINE   209
The Art of Compassionate Discipline   216
Take Ownership   218
Be More Like Usain Bolt   220
Progress Is Greater Than Perfection   222

A Few Final Words   227

Appendix: Welcome to Your 30-Day Visualisation
   Challenge   232
Resources   244
Acknowledgements   245

INTRODUCTION

Work on your mind like you work on your body –
the strongest version of you depends on both.

Growing up, physical education was my favourite lesson in school. From the age of four, it was ballet or playing cat and mouse in the sports hall. As I got older, it was bench ball, tennis and athletics. By the time I was in the sixth form, I was playing five sports and going to the gym nearly every day. I grew up understanding the importance of taking care of my body and exercising to stay physically fit. Even now, I continue to prioritise playing a sport, going for walks and spending time at the gym. It's a normal and necessary part of my life. And, when I don't, I immediately notice the toll it takes. Many people are the same. Even though not everyone is a gym fanatic or expert runner, most people know that they should prioritise moving their bodies.

PE was a compulsory part of the school curriculum, and yet, no one ever mentioned the importance of training our minds. Did we attend ME classes? Mental education?

Did we ever learn how to train the brain? Our mind is half our health. From the moment we wake up, it dictates everything we do – the way we think, the way we feel, the way we perform – so if we don't train it or know how to take care of it, how are we meant to stay mentally fit? As a society, we put so much importance on our bodies and how we look. This translates into the way we show up in our relationships as well as how we prioritise our habits. It makes sense – the body is visible and therefore we can see the results. But do you have power over your thoughts? Can you manage your emotions? Do you know how to rewire your brain to become more confident? Or what about achieving your goals without burning out? Nearly everyone is struggling with aspects of their mind, but most people aren't empowered with the tools to address it. Training the mind has been completely neglected and we now suffer the consequences of that to varying degrees. No matter how old you are or what you do, your mental fitness is crucial to a healthy, fulfilled and purposeful life.

You might have noticed – I am not using the term 'mental health'. This is intentional. Everyone has mental health. If you have a mind, you have mental health. Just like if you have a body, you have physical health. But not everyone is mentally *fit*. You stay physically fit by lifting weights, running or perhaps doing pilates. In the same way, you are only mentally fit if you are actively training your mind with the correct exercises and routines. It's time to stop only talking about mental health and instead actually do something about it. That's why I specifically talk about 'mental fitness'.

The term was first used in psychological literature in 1964, and yet still very few people know about it, let alone practise it. In fact, I recently took to the streets of London and asked strangers to rate their mental fitness from 1 to 10; 90 per cent of people looked at me blankly and answered with 'What's mental fitness?' Swapping the word 'health' with 'fitness' may seem like a tiny change, but it makes a big difference. Mental fitness encompasses the dynamic nature of our minds and how we can take intentional action to maintain good mental health. It is more empowering, action-oriented and sustainable. While mental fitness is an umbrella term for many things, this book will unveil a technique with the potential to profoundly impact your life: visualisation. Visualisation didn't just change my life – it saved it. But we'll get to that in a moment. The technique has allowed me to achieve my goals, build confidence and rewire my brain. I wish I had learned it sooner.

I went from being:

— Limited by insecurities like fear of failure, self-doubt and the need for external validation
— Easily overwhelmed by challenges and quick to give up under pressure
— Defined by feelings of inadequacy, self-criticism and a constant need to prove my value to others
— Chronically stressed and consumed by my emotions, with little ability to regain control or perspective
— Quick to sacrifice my own needs and say yes to everything
— Drawn to unhealthy dynamics and relationships that drained me emotionally
— Overcommitted to tasks and responsibilities, leaving no time for rest or personal growth
— Afraid to step into my potential

To someone who now:

- Knows their value, sets high standards and refuses to settle for less
- Is equipped with resilience, clarity and emotional balance to handle challenges with composure
- Is confident in their abilities, embracing risks and opportunities with conviction and determination
- Is attuned to their inner signals, prioritising rest, self-care and mindful decision-making
- Is thriving with vibrant energy, a positive outlook and a deep sense of joy in daily life
- Starts each day with appreciation, seeing opportunities and beauty in even the smallest moments
- Strategically approaches challenges, achieving success through sharp insight and relentless effort
- Lives authentically, unburdened by limitations, and is deeply satisfied with their purpose and journey
- Approaches life with boldness, trusting in their ability to overcome obstacles and achieve greatness

What changed? Sure, there are many things, but definitely the most profound is that I practised visualisation for over six years (and still do). Investing in my mental fitness allowed me to improve the way I think, feel and perform – and it could do exactly the same thing for you.

INTRODUCTION

THE FIVE KEY AREAS OF MENTAL FITNESS

People overcomplicate mental fitness. I've spent the last five years uncomplicating it for myself and now I want to share it with you. In my recent TEDx Talk, I broke down mental fitness into five key areas: **consistency**, **cardio**, **diet**, **rest** and **weightlifting** – the same principles we use when working on our physical fitness. Using this analogy makes mental fitness tangible and achievable. Let's explore these:

— **Consistency:** How often do you train your brain? You don't get abs of steel overnight. The brain works in the same way. It requires regular reps. It's not about being extreme, it's about being consistent.
— **Cardio:** If you don't use it, you'll lose it. Activities like doing puzzles or quizzes, journalling, critical thinking, learning and debating proactively engage the brain to keep it sharp.
— **Diet:** Your diet isn't just about the food you eat, it's about the content you watch and listen to, including the conversations you have. You want to feed your mind with the good stuff! Take a moment right now and think about the content you read, watch and listen to. Is it nourishing your mind or draining it?
— **Rest:** How do you give your mind a break? Whether it's through sleep, meditation or a digital detox, regular and active rest is crucial for mental

> recovery. Just as your muscles need time to repair after a workout, your brain needs downtime to process and recharge. Ask yourself: when was the last time your mind actually felt rested?
> — **Weightlifting:** This is about building flexibility, strength and resilience in the brain. How can you upgrade your mindset to handle life better and achieve your goals? That's where visualisation comes in.

Contrary to what you might think or experience at times, your mind is on your side. And the more you strengthen it, the more that will be the case. While all the different aspects of mental fitness are important, I urge you not to neglect 'weightlifting'. Visualisation in particular helps you to unlock a depth of potential you may not even know exists. As I have seen from my own experience and others', your relationship with your thoughts, character and performance undergoes a complete transformation. Committing to strengthening my mind gave me a mental edge – one that helped me not only achieve my goals but also overcome the obstacles in my path. That doesn't mean I don't still face tough days or difficult periods. But now, I face them with a strength, clarity and resilience I didn't have before. And that's the true essence of mental fitness.

One thing I want to stress is that you don't need to be unhappy, struggling or unwell to practise visualisation or read this book. This is a dangerous assumption. Do you only go to the gym when you are injured or overweight? Or do you go because you want to stay healthy, remain fit and feel great? Treat your mind in the same way. Instead of only taking action when you have to, take it when you can and want to. That's what champions do.

MY STORY

Today, I am the UK's leading mental fitness and visualisation expert. Day to day, I work with global brands, as well as delivering events and keynotes internationally. I coach elite athletes, executives and business leaders – from top Premier League footballers and world-class tennis players to billionaires, CEOs and influential public figures. Of course, my story didn't begin like this!

Growing up, I was what you might call a 'high achiever'. I excelled academically, danced competitively and played professional badminton. As a South Asian, the expectations were always high, and I was used to meeting them. I was confident, outgoing and had a reputation for being social and bubbly. People often told me I had endless potential, and I believed them. It made me excited about my future, feeling like the world was my oyster. But beneath the surface, I was also highly sensitive. I felt everything deeply, and this led to friction in my relationships. Jealousy crept in, especially in competitive environments, and the pressure of sports often weighed heavily on me. In hindsight, I realise that while I appeared confident on the outside, I was actually quite insecure on the inside. Despite this, life was still pretty amazing. I had a supportive family, loved school, and had the privilege of training five or six times a week while competing in tournaments across the country. All in all, I was happy.

In December 2013, at the age of 15, I started experiencing constipation and saw streaks of blood in my stool. It was a little scary, but I didn't think too much of it – perhaps it was some sort of stomach bug. I also wasn't feeling much pain so didn't think it was too serious. I went to the doctor soon after and was prescribed some medication. Things didn't improve much, and, after getting a colonoscopy, I was eventually diagnosed with ulcerative colitis. Colitis is an incurable inflammatory bowel disease. I was told that this was quite a common disease among 15–30-year-olds and that I would have periods in my life when I would be ill, which are called 'flares', and then periods when I would be healthy, which are called 'remission'. The good thing was that a lot of people can have ulcerative colitis and live a normal life because there is medication that can manage the disease extremely well. I'd never heard of the disease before, but aimed to be positive, followed medical instructions and continued life as usual. Unfortunately, as time wore on, my symptoms got worse and the reality of the disease began to set in.

A flare would start gradually, almost deceptively. At first, I'd notice some discomfort after meals, a cramp or ache here and there, but within days, the symptoms would escalate. I would find myself running to the toilet up to 30 times a day, always with bleeding, intense urgency and a pain so sharp it felt like my insides were being scraped. At night, I woke up every hour drenched in sweat, clutching my stomach, screaming into my pillow from the pain. Sometimes, I even had to sleep on the loo because, after experiencing the excruciating pain, I didn't have the capability to walk even a few steps back to my bed. Even writing this now makes me feel emotional

INTRODUCTION

because I remember how inhumane it all felt. Every part of my body felt fragile. My stomach was so sensitive that even the idea of eating or drinking filled me with fear, knowing the pain that would follow. I watched myself shrink in the mirror as my weight dropped alarmingly fast and my face became gaunt.

But flares weren't just physically excruciating. The disease took over my mind as well. I was in a constant state of fear and anxiety as well as experiencing intense shame around the symptoms. So much so that no one other than my family knew about them. In hindsight I wish I had told someone, but being a young teenage girl, this wasn't something I felt comfortable to speak or open up about. It was an isolating experience going through this for nearly five years.

During a flare, the only thing that gave me some relief was steroids. They didn't solve the problem but they made the pain more bearable and reduced the symptoms. Slowly, they gave me a chance to rebuild, gain a little weight, go back to school and try to act like a regular teenager. But I was far from it. At my worst, I was going to the loo up to 40 times a day and taking 65 tablets a day. I would go to school wearing adult nappies, running out of classes, having accidents as I was walking around and having to hide my pain from everyone.

I hated what my life had become. My world revolved around pain, medications and hospital visits. I started to resent my friends and even my family. I envied their laughter, their freedom, their normality. Watching people eat casually, play sports or make plans for the future made me feel bitter and alone. I wasn't just in pain; I was broken. Going through one flare was torturous enough, but imagine going through this nine times. That's three years of constant agony.

With each flare, my mind and body were becoming weaker and weaker. I remember lying in hospital one day and just feeling defeated. In fact, the word 'defeated' doesn't do it justice. Physically, I was a shell of a human. I was 17 and in the midst of one of the worst flares of my life. I had been hooked up to IV steroids and painkillers for days. There was nothing behind my eyes. I said nothing. I felt nothing. I'd lost everything. Every part of my body and mind had given up. I was so broken from the repeated pain and symptoms – I had no fight left in me. Many people would tell me how strong I was for enduring the symptoms, but I was so tired of being strong. And, truthfully, I didn't want to fight anymore. It was so scary going from a young girl with so much passion and fire for life to someone who had lost everything that made me feel human. I kept asking myself why I was being punished and why this was happening to me. I'd lost all hope. I couldn't see any way out.

A few days later, my doctor recommended surgery. My body wasn't responding to the medication, and urgent action was needed. My colon was at a high risk of rupturing, which there would be no turning back from. The procedure would involve removing my large intestine and attaching a stoma bag to my stomach. I remember bursting into tears, feeling completely overwhelmed. How was this the only option? I was just a teenager, and here I was, being asked to make such a huge, life-altering decision. The medical team was confident it was the right choice, but for some reason, everything in me was screaming no. I asked for some more time, even though I really didn't have it.

INTRODUCTION

One morning in hospital, the nurse who took my bloods each day came into my room. She could see I had been crying all night. To try to help, she asked me what I would be doing if I wasn't lying in this room. I barely looked at her. In my mind I was cursing her because I found it to be a rude and thoughtless question. Here I was, stuck, with no hope for my future, and she was asking questions that made me feel even worse. I asked her to leave my room. When she left, however, I remember slowly closing my eyes and thinking about what she had said. The first thing I wanted to do was walk again. I hadn't left my bed in about a week and wanted to feel just slightly human. So, I started to build an image in my mind of myself walking down the hospital hallway, slowly but steadily. When I opened my eyes, I took a second and felt slightly weird. There was this tiny part of me that said, 'I want to see myself doing that again', and then I remembered where I was and faced my reality. For some reason, though, I kept thinking about the image I saw, so I closed my eyes once more and continued to watch myself walking. This time, I spent longer doing it, making the picture even more vivid – seeing the other patients who passed me, waving to the nurses, literally feeling each step. My brain naturally started pushing me a little further. I saw myself walking near my house and then running. I then started seeing myself healthy again – laughing, learning and being with my friends. It felt real.

Each time I did this, I started to feel a little bit of hope. After hearing again and again that I would be ill for the rest of my life and that I would need surgery, this image I kept rehearsing showed me another possibility. I decided to make it my goal. Every spare minute I had in that next week in hospital, I showed my mind what walking and running looked and felt like again. The more I did this, the more I felt the confidence to get out of bed and take just a few steps. Even though I was sometimes still in pain and feeling weak, I knew what I wanted to achieve. And I did it! I had made something that seemed so out of reach happen. The week after, I'd gained enough strength to at least be discharged from the hospital and return home to recover.

I was still too unwell to go to university and had to spend the following year recovering from one of the most painful flares I had ever had. My bones were so weak from all the steroids. I was about 37kg (5st 12lb) at the age of 18 and I didn't know what it meant to live a normal life anymore. And although it may seem like a small problem in the context of what I am explaining, it was a hard reality to swallow seeing all my friends leave school and start their new lives, while I was battling what felt like life or death. I was still getting pressured to go ahead with the surgery as that was the only option left.

In my recovery time, I dedicated everything to learning about why mentally rehearsing myself walking had helped me to get stronger, how the mind works and all the ways I could keep getting better. I read every book I could find, listened to stories – some were even on cassettes! – did courses, trained with neuroscientists and, when I'd gained enough strength, attended some retreats. While all of this was going on, my parents and I were trying to find every possible solution to help me. We met with doctors abroad and tried their treatments. I went on every diet that claimed it would help, from only having sunflower seeds and spinach and litres of celery

INTRODUCTION

juice, to grain-free and liquid-only diets. I was exploring hypnotherapy, acupuncture, Chinese herbs and meditation. I even looked into getting a faecal transplant – something that was barely legal in the UK! It was a constant fight for survival. But something that I always kept coming back to was visualisation, partly because it was a huge source of hope for me but also because I was noticing improvements.

It was around this time that I was introduced to the kindest woman I know, who soon became my visualisation teacher. She had been through her own struggles in life. The first time I went to her house, I sat down and cried for an hour. The second time, I did exactly the same thing. I didn't have anything to say – because I had been in survival mode for so long, I had so much pain and emotion to let out and this was the first time I felt safe enough to do so. After five sessions, I finally said something and started processing what had happened in the past few years. Within this work, she taught me how to expand my visualisation practice. I visualised my colon healing in immense detail. I visualised going to university and saw myself in the Bristol University badminton kit. I visualised eating solid food again and enjoying it. I visualised what a healthy Maya looked like. Five or six months later, I started to let myself think even bigger: what could my longer-term future look like? I mentally rehearsed building a company and speaking on stages. Day in and day out, I was revisiting my goals and healing my body with the aim of rewiring my mind for a better future. This included altering old beliefs I held about myself and addressing negative thought patterns. I worked through emotions I had been suppressing, my memories of being ill and my future potential. And that's when I really discovered the power of visualisation – the ability to rewire the brain, heal my past and take control of my future.

Writing this out makes it sound so simple, but it definitely wasn't an overnight success. In reality, that year was gruelling. I still experienced intense fear of flaring again or things not working out. I had to face some of the darkest parts of my personality and I was still dealing with difficult symptoms at times. There was no guarantee that there would be glory on the other side of it, so naturally old patterns kept coming up. It required a new level of discipline and resilience that no one can ever teach you. To push on every day despite what my blood test results were saying or sometimes even how I was feeling demanded so much mental and physical strength.

After a year of rigorous work on myself and starting a new medication, I was lucky enough to attend the University of Bristol, studying Geography with Innovation. I religiously practised my visualisations before my lectures each day. As I started to feel better and return to normal life, my visualisations focused on becoming more confident, graduating and building a business. I kept learning about the mind. I kept researching how to train it, but, most importantly, I was practising and getting results.

Going to university felt like a dream. I was now living out the vision I had kept rehearsing of where I wanted to study. At that point, I was finally no longer experiencing symptoms or flares and was in the process of rebuilding my life. But during my third year, something shifted. I started noticing the familiar tightness in my stomach again. It grew uncomfortable, and one day, I saw blood in my stool. I had been under the impression that I had overcome this illness, but that moment shattered everything. The thought that I might have to endure years of crippling suffering again was terrifying. I felt angry, disheartened and like a fraud.

It took me some time to accept what had happened. But then this tiny voice in my head kept saying that if I could do it once, I could do it again. In fact, I *had* to do it again. I'd had a taste of what my life could be and that gave me purpose. I once again drew upon all the tools I had learned and reminded myself that my mind is the most powerful asset I have. So, no matter what I was faced with, I was going to harness the power of visualisation to get my life back *again*.

During this period I put everything I had into strengthening both my mind and body. I worked closely with my visualisation teacher, deepening my practice and confronting suppressed emotions of shame and sadness. I had to work hard to rewire my beliefs about the disease, what it took from me and what it meant for my life. Alongside diet changes and significantly reducing stress, the key was training myself to truly believe that I could overcome this and visualising my recovery no matter what anyone else said. Nothing happened overnight, and trust me, I wanted to give up many times. But it was all worth it. Because through this work, I have gone from strength to strength and have lived the last four years of my life medication-free, flare-free and the happiest I have ever been.

While I started my journey into mental fitness through illness and disease, I continue it because I want to achieve greatness in my life. But I am going to be straight with you: I am not a neuroscientist, nor do I have a PhD in sports psychology. I am also not a doctor! Everything I know, everything I teach and everything I have built comes from over a decade of real-life experience, learning and results. Oh, and a slight obsession with everything about the brain. My mission is simple: I am making mental fitness and visualisation as common and as accessible as physical fitness. Because here is the truth – people who have had the privilege of being taught visualisation and have applied it to their career, life and character have been living with an unfair advantage. I am one of those people and, in this book, I am going to let you in on all the secrets, techniques and ways in which you, too, can build unshakeable mental fitness.

HOW TO USE THIS BOOK

Visualise is written in three parts: Awareness, Intelligence and Execution.

Awareness lays out the foundations of how the mind works. This is level one of mental fitness: learning the power of your thoughts, emotions and beliefs, but, more importantly, the brain's phenomenal ability to rewire. This section will also outline the foundations for setting goals and pursuing them.

Intelligence, level two, is probably my favourite section. It's all the practical guidance I wish I'd had when I started learning visualisation, and compiles seven years' worth of practice, research and case studies, which aren't documented anywhere else. This level is for the action-takers. The exercises, questions and visualisations in this section are based in neuroscience and sports psychology, and are backed up by years of tried-and-tested results.

Level three – Execution – is all about taking what you have learned and implementing it into your life for the long term. This level requires consistency, discipline and commitment – things that people in modern society often struggle with. But it's the difference between feeling mentally fit for one day versus acquiring the skills to stay mentally fit for life.

When you finish reading this book, I hope you will have gained a mental fitness education, something we should have had growing up. With the tools you learn, you can immediately start implementing these techniques and improve the way you think, feel and perform. To make it even easier, I have designed a 30-day challenge (Appendix, page 232) to help you build the habit of visualisation and start making meaningful changes in your life.

Some of you will read this book and never look at it again. Others will read this book and study it. You will make notes, you will try the exercises, you will challenge it, you will be focused. Keep an open mind, stay curious and have fun! And never underestimate the power of one. All it takes is that one sentence, one quote, one visualisation or one question to change your life.

Before we get started, I only have one more thing to say:

**Your mind is like a puppy.
If you don't train it,
it will sh*t everywhere!**

PART 1

AWARENESS: LEARNING THE FOUNDATIONS

The mind is both the key and the lock to every door we wish to open.

A few years ago, I was listening to Sir Ian McGeechan on *The High Performance Podcast*. He spoke about the importance of getting the 'world-class basics right'. He continued to explain that the best athletes spend time learning and relearning the foundations. This might be basic footwork, basic skills or even basic strategies. The same applies to mental fitness. No matter how complex and advanced techniques can become, nothing beats the basics. Understanding the foundations that underpin our mind is the first step to becoming mentally fit. Some of you may already be familiar with these aspects, while, for some of you, they might be new.

As you read through Part 1, stay curious about your own mind and patterns along the way.

CHAPTER 1

HOW VISUALISATION CAN REWIRE YOUR BRAIN

You are not stuck with the brain you were born with. You can teach an old brain new tricks.
– Dr Norman Doidge

The world of personal development is full of different tools, theories and practices to help the mind and body – from meditation, journalling and cold therapy to neurolinguistic programming (NLP), tapping and breathwork. I practise a lot of these, and have done for years. But there is one technique that is largely unknown or less understood and I would argue is one of the most powerful mental training techniques there is. Visualisation (also known as mental imagery or mental rehearsal) is a specialised mental process of using your imagination to create vivid images, environments and feelings in your mind and body before they happen. The aim is to engage all your senses and intentionally manufacture your imagination to rewire how the brain works – influencing your thoughts, emotions and actions. Visualisation can be traced back to the likes of Aristotle and Zeno of Citium in Greek philosophy and today is most used in professional sports, advocated by the likes of Michael Phelps, Serena Williams and Mo Salah. For example, the boxing legend Anthony Joshua mentally rehearses each fight hundreds of times over in his mind before going out. This includes his walk-on, the roar of the crowd, his opponent's moves and his own performance. He even blasts crowd noises during his training sessions to fully immerse himself in the experience. By the day of the actual fight, he has gained significant confidence, focus, muscle memory and resilience. Due to its success among athletes, visualisation is now being used within other high-performance settings, such as medical and surgical training, music, aerospace, rehabilitation and the special forces.

More recently, it has been popularised by renowned experts like Joe Dispenza, David Hamilton and Tara Swart. Dispenza has made some incredible discoveries about the power of visualisation for overcoming barriers and changing your reality. I followed a lot of his work when I was recovering because of how convincing the results were. I remember practising his visualisations every morning and night, consistently seeing myself as healthy and healed. In her book *The Source*, Tara Swart speaks with authority and practicality about using visualisation for achieving goals and success. But it still surprises me how few people know about visualisation. And, more specifically, know how to use it in their life as a mental fitness tool. After years of working in this space, I've come to realise that the barriers are a lack of education or understanding and, at times, even gatekeeping. Elite performers have always had access to the best psychological tools money can buy, including visualisation, but they don't always break into mainstream culture. Even now, many up-and-coming athletes aren't aware of the full potential of visualisation. And when I speak to former athletes who've used the technique to achieve their sports goals, they often don't know how to transfer those skills to other areas of their life, like public speaking, relationships or emotional management. While I am not an athlete aiming for the Olympics, I have my own version: the company I'm building. So, if these tools can work for them, why can't they work for all of us? People are missing out on something that could significantly alter their life. It's time to make this knowledge accessible to more people. Whether you are an elite athlete, business owner, artist, student or parent, this is a tool that can change your life for the better.

We'll look in detail at the five visualisation techniques in Chapter 5, but first I want to explore the amazing way in which our brain can rewire itself, and how visualisation can strengthen new neural pathways in the brain. Once you understand the science behind visualisation and how adaptive the brain is, you'll gain a deeper understanding of how your mind works.

NEUROSCIENCE 101

Like many people, I was skeptical of visualisation at first and didn't know about it as a tool I could use in my day-to-day life – that is, until I learned the neuroscience behind it. There are four different but interconnected ideas worth learning about as they form the basis of how you can rewire the brain.

Neuroplasticity

Let's start with neuroplasticity – the brain's ability to change, rewire and repair itself. For a long time, it was believed that the brain you have in your early twenties is the brain you will always have.

However, neuroscientists have now discovered that the brain is far more adaptable. It can form new connections between neurons, shifting your brain's responses, habits and ways of thinking. This leads to lasting changes in behaviour, emotions and cognitive functions.

While it's impossible to grow new neurons, we can create new *connections* between them. The human brain has around 86 billion neurons, forming nearly 100 trillion connections, or neural pathways. Every time you think, feel or act, these pathways light up, firing electrical impulses. They shape our habits and patterns of behaviour. For example, brushing your teeth twice a day is a deeply ingrained habit with a very strong neural pathway. It's so strong that breaking it would be harder than sticking with it.

According to neuroscientist Tara Swart, rewiring the brain becomes more challenging after the age of 25. When we're younger, our brain is still growing and organising its structure, making it more flexible and open to change. As we age, however, our ways of thinking and established neural pathways become more rigid. But the great thing is, neuroplasticity is still possible. The key is to be more aggressive and intentional with the practices you adopt. There are many ways we can make the brain more neuroplastic. For example, learning a new language or skill creates new pathways. Or brushing your teeth with your less dominant hand forces the brain to engage different neural pathways to perform a familiar task. In both cases, you would start to carve out a new road in the brain. And if you keep travelling along this road, the brain will begin to use this pathway more. This then becomes the new habit. The old pathway gets used less and less, and eventually may even disappear.

But there's another way to rewire the brain besides physically doing or learning something different – visualisation. To really understand how, it's important to grasp one important point: **your brain struggles to tell the difference between what is real and what is imagined**. When you physically perform an action like playing tennis or speaking in public, certain neurons in the brain fire, activating specific regions. When you mentally rehearse the same action, nearly identical pathways in the brain activate. Put bluntly, as clever and remarkable as your brain is, it's actually pretty simplistic. The mind really struggles to differentiate between vivid mental imagery and physically doing something. For example, when you close your eyes and visualise yourself walking into a room confidently, you're not just imagining it, you're training your brain to experience it as real. Repeatedly doing this triggers neural pathways that reinforce self-assurance. Over time, this mental rehearsal can actually reshape the brain, making it easier for you to act with confidence in real life. This neural reprogramming translates into tangible improvements in how you carry yourself. That's why visualisation is such a powerful tool. Each of the five techniques we'll look at in Part 2 works because they take advantage of how you can harness your imagination to create change. It's why my client Kallie can manage anxious feelings in 90 seconds. It's why James, a tech entrepreneur I work with, is becoming a more confident public speaker and communicator. It's why Sally, who came to one of my events, decided to leave her job and become a yoga instructor. It's why I can achieve the goals I set my mind to and not burn out. It's also the reason why so many of us are stuck repeating the same patterns and getting the same results.

While neuroplasticity is happening all the time in our brains, it's important to note that too much stress – whether that's internal or external stress – can significantly reduce the brain's ability to form new connections. The great thing about visualisation is that, by relaxing the mind and body beforehand and then carrying out the mental imagery, you can take full advantage of the brain's incredible ability to rewire itself.

VISUALISATION IS WEIGHTLIFTING FOR THE BRAIN

When we physically lift weights, the muscles in our body tear and then get rebuilt. It's a way to strengthen the body and prevent further injury. As you get older, it's also a way to maintain bodily function and muscle mass. I liken visualisation to weightlifting because the technique involves breaking down neurological patterns to build new connections and strengthen existing ones. The ability to continually rewire the brain means you

> remain more agile and strong in challenging situations while also pursuing excellence. It's the key to building sustainable mental strength and fitness.
>
> For example, I'm currently using visualisation to change how I respond to feedback. In the past, I'd take feedback extremely personally, letting it crush my self-esteem and halt my progress. Now, I visualise myself calmly receiving feedback. I imagine staying composed, breathing deeply and listening without reacting. I mentally rehearse various situations where I respond with curiosity rather than defensiveness. Recently, when I received some feedback, I was able to step back, take it in and use it to improve my work. While I still have room to grow, I'm already noticing positive shifts in my reactions.

Mirror neurons

As well as struggling to tell the difference between what is real and what is imagined, the brain finds it hard to distinguish between whether you are carrying out an action or watching someone else do it. **When an individual observes an action, their mirror neurons activate as if they were performing the action themselves.** And if they have previously carried out the action themselves, the neurons are further activated. It's why I tell athletes I work with to watch and rewatch videos of themselves or others performing extremely well. The mirroring process enables individuals to understand and internalise the actions, emotions and intentions of others, essentially allowing them to improve their performance and results by observation.

But this can also work in the opposite way. Growing up, I used to watch reality TV shows. Subconsciously, I was learning a lot of unhelpful behaviours, such as negativity, conflict and manufactured drama, that would subtly show up in my own life or relationships. I was mirroring the things I was seeing. The more I fed my mind with them, the more I acted in that way. At first, I wasn't aware of it. Once I learned how the brain works, it all made sense! Sometimes it's as simple as observing what our parents do or how our friends act. That's why we must be extra careful with what or who we watch.

The brain's filter system

The brain cannot pay attention to everything. But what you pay attention to will either expand or contract your worldview. And that's why we have a reticular activating system (RAS). This is a collection of brain areas that decides what information is brought into your awareness. I like to think of it as the sieve of my mind. We consciously process about 50–100 bits of information per second. The RAS works by filtering out what it thinks is irrelevant information. The red car theory is often used to showcase the RAS in action. Let's say you wanted to buy a new red car. You will start to notice the exact car you want everywhere you go. Even though its prevalence hasn't changed, because you are now focusing on it, you are more likely to notice it. In the same way, because I visualise my goals every day – whether that is me walking again, being more confident or building a global company – I am instructing my RAS to pay attention to the things, people and opportunities that will help me achieve my vision. It directs my focus and attention to that goal. **While we consciously create the vision, the RAS works subconsciously to identify pathways towards it.** When these pathways lead to successes, it may feel like luck or coincidence, but it is not luck at all. It began with a vision we created and actions we took based on the information filtered by our RAS.

Neurons that fire together, wire together

In 1949, Donald Hebb, a Canadian psychologist, postulated that when neurons fire together, they wire together. For example, when you repeat an experience over and over, such as playing a musical instrument or learning how to drive, the connections between activated brain cells become faster, stronger and more efficient. They get deeply engrained into the brain. **Experiences that are intense, prolonged or repeated can change the brain.** Remembering the earlier concept that the brain really struggles to know the difference between mentally rehearsing something and physically rehearsing something, Hebb's law becomes even more interesting in the context of visualisation. Repeated mental rehearsal of better performance, improved confidence and achieving goals will initiate the new firings between neurons, eventually leading to new wirings. But this can sometimes work in the opposite way; we can rewire the brain to think more negatively. For evolutionary reasons, the brain reacts more to negative experiences than positive. That's why we have to be intentional with what we visualise.

I've experienced the benefits of new wirings many times. For example, I used to be afraid to speak on stage in front of people. One time, I had an accident on stage due to being in a flare-up with colitis. The experience was intense and emotional. From that moment, my brain was wired to be fearful of public speaking. But, since then, I have visualised speaking on stage hundreds of times. Each time I do this, I am firing new thought patterns that can wire into new beliefs and better skills. I now speak on stages all over the world in front of thousands of people. You, too, can harness the power of neuroplasticity through visualisation to build new pathways.

THE BRAIN IS LIKE A CITY

I like to use the analogy of your brain being like a city to explain how you can rewire it to change old patterns that are holding you back.

Imagine:

Your thoughts = cars
Your beliefs = roads
Your experiences = buildings
Your emotions = the weather

Every second, you are driving along different roads and constructing new buildings. The more you drive on a road, the more permanent it becomes. These roads are then your mental map or blueprint, which is how you navigate the world. It becomes the architecture of your mind. Some of the roads will be motorways, others like country roads.

HOW VISUALISATION CAN REWIRE YOUR BRAIN

Some were built by your parents or teachers. Naturally, you will drive most often along the roads that are most prominent or that you are more familiar with. Sometimes, there will be traffic jams or car crashes. Sometimes, the buildings will be ugly. And sometimes, there will be no road in front of you. At the same time, some days it will be sunny, on others it will rain and thunder. The point is, there is a lot going on, but it's our job to take ownership and build the city the way we want to. To do that, we must learn how to break down old roads (beliefs) and build new ones. Where things like meditation and mindfulness can help us slow the traffic, visualisation changes the roads in our brain to roads that serve us – roads that help us achieve our goals, build our confidence and increase our resilience. This is about you becoming the architect of your brain.

When I speak to people after events and keynotes, or even to my clients, they always tell me they wish they had known about visualisation earlier or had the right information to start practising it. But everything you need to be good at visualisation, you already have. In fact, most people are already experts at it. The problem is, we visualise the worst that is going to happen. We play out scenarios of us failing or things going wrong. We drive along old roads and experiences that don't serve us. I used to do exactly the same. In the process, I was wiring my brain to be more anxious and insecure. But if you want to enter a new chapter in your life, you've got to stop rereading old ones. Whether you are starting something new, making a big change in your life, recovering from a hard time or trying to reach new heights, your mind is your biggest asset. Practising visualisation is about learning to use your mental capacities in a productive way, in effect choosing the way you want to think, feel and perform rather than leaving it to chance or conditioning.

The quality of your life is a direct reflection of the quality of your thoughts, emotions and beliefs, so it's time to step up and build your own city.

CHAPTER 2

PAY ATTENTION TO YOUR THOUGHTS

Your thoughts are either weapons or tools.
You get to choose if they harm you or help you.

Let me begin this chapter with my favourite story. It's an old Cherokee legend about two wolves. A grandfather explains to his warrior grandson that there are two wolves within each of us. They are constantly fighting for control over our mind. One wolf is positive and encouraging, while the other is negative and destructive. The grandson asks, 'Which wolf will win?' The grandfather replies, 'The one that you feed.'

On average, a person has between 20,000 and 60,000 thoughts a day. That's a lot. Some thoughts are fleeting and harmless, like what to eat for dinner. Some are more creative, like an idea for a new business. Others are more critical and louder, like telling yourself you aren't good enough or thinking, 'I don't deserve this success.'

While it's impossible to track every thought you have, there are some we want to – and can – pay more attention to. A single thought in and of itself doesn't make too much of a difference, but the repetition and nature of the thought have a big role to play. If you are constantly having negative thoughts about yourself or the world, that becomes part of your identity. 'What if this fails?', 'I'm an anxious person', 'I'm not sure I can win', 'People don't like me' – these are all going to have a huge impact on your confidence, pursuit of goals and relationships. It's just like when we create physical habits. When we repeat an activity, we create a habit and the task gets easier. Similarly, the repetition of thoughts creates mental habits and it becomes easier to think in that way. This type of thinking can either be detrimental and pull us down or lift us up and serve us well.

You might have heard of the phrase 'your thoughts create your reality'. While this shouts 'Instagram quote' from a mile away, the statement remains true. Two people can have the exact same experience but have different thoughts about it. For example, let's say Alice and Joel are both in the same traffic jam:

Alice: 'This is so frustrating! I'm wasting my time, and now I'll be late. This always happens to me!'

Joel: 'I can't control the traffic, so I'll just relax and listen to my favourite podcast. I'll get there when I get there.'

Even though they are both in the same situation, Alice's negative thoughts create stress and frustration, while Joel is able to remain calm and content. It all comes down to how a situation is perceived:

How you *think* influences how you *feel*.
↓
How you *feel* influences how you *act*.
↓
How you *act* determines your *behaviours*.
↓
And your *behaviour* becomes your *identity* and *reality*.

THE MYTH OF 'ALWAYS THINK POSITIVE'

It might sound like I am about to contradict what I have just written, but hear me out. When I first started learning about the power of my thoughts, every book would tell me that my thoughts should always be positive. And if they weren't, I'd never be happy or healthy. There was no room to think negatively otherwise I would remain sick for the rest of my life. Theoretically, I understood why, so that's exactly the advice I followed. In every situation, I forced my mind to stay positive, especially as I was trying to overcome my illness. I found it extremely hard. But the interesting thing I started to notice was that if a negative thought came up, I felt guilty for having it and scared of the damage it would do to my life and body. The 'shoulds' and 'oughts' consumed my mind and, in my pursuit of trying to think more positively, I ended up stuck, criticising myself for not having positive thoughts all the time. Ironic, right? My relationship with my mind became fearful. I remember reading that it takes three positive thoughts to negate a negative one. I took this statement so literally that I ended up forcing myself to counteract any negative thought. Quite honestly, it was draining, and it felt like I was playing a game I couldn't win.

There is an increasing pressure in society to think in this way. Lose your job: 'positive vibes only'. Break up with someone: 'don't think about it, just move on'. Scared about your future: 'always look on the bright side'. Sometimes it does more harm than good because we end up denying what we are actually feeling. Thinking happy thoughts doesn't just eliminate what we are going through. Don't get me wrong, it can definitely help in some situations, but forcing positivity isn't always the answer. We must explore a deeper relationship with our thoughts and minds.

The more I studied the mind, I realised that the aim is not to have positive thoughts all the time. Because, quite frankly, that is impossible and not natural. Rather than the pressure of only having positive thoughts about yourself and the world, shift it to having a positive *relationship* with your thoughts: one where you have control. It's like any other relationship in life – romantic, friends, family and so on – they aren't always going to be positive, but the key is being able to stay dynamic and flexible. We don't need to fear our thoughts, specifically the negative ones – it just comes down to what you think about *most* of the time. That's where the power lies.

Being an expert in mental fitness doesn't mean I don't have negative thoughts. I also feed the bad wolf and that's okay – I'm human. The difference is that now those thoughts don't consume me. I am able to recognise them, to let them pass without judgement, and, if needed, I have the tools to change them. It's not always an easy task, but it's possible. Just the other week, I had a conversation with a client who is crushing it in her career. But in the middle of our chat, she started saying, 'I'm not working hard enough. I'm falling behind.' The thoughts made her feel inadequate, but because she's been working on her mental fitness, she was able to catch them. She reminded herself that it's normal to have such thoughts and feelings and that everyone is operating at their own pace. Unlike when I first met her, she didn't let those thoughts consume her, and instead, she got back to focusing on her goals. In other words, you can't get rid of all of the wolves in the brain, but you can learn how to tame them.

Dr Price Pritchett recorded a fascinating episode with Lewis Howes on *The School of Greatness* podcast. In it, he brought up a question that he asks the audience at his events: 'What do you think is more important? More positive thinking or less negative thinking?' I didn't know the answer. As he went on to explain, the studies are 'unequivocal'; positive thinking is important, but being able to reduce negative thinking is what makes the most difference to your mindset and overall fulfilment. But remember, this doesn't mean the aim is to go through life without a single negative thought – the aim is to change your response to them. The Indian spiritual teacher Sadhguru goes as far as saying that we should refrain from labelling thoughts as positive or negative. Instead, just see them as thoughts. I like this idea because it encourages us to remove unnecessary judgement.

So, if you were like me and concluded that the answer to a strong mind is to think positive all the time, I urge you to think again. Instead of trying to completely flip the scale and have 100 per cent positive thoughts, start asking yourself, how can I stop feeding the bad wolf and feed the good wolf a little bit more? How can I have a negative thought and not be consumed by it? I would suggest starting with one or two thoughts you repetitively have that don't serve you, and challenge yourself to either change them or respond to them better. This is where visualisation comes into its own. Through neuroplasticity, we can reprogram the subconscious mind by unwiring negative thought patterns and strengthening new pathways. We'll be looking at the different ways you can do this in Part 2. With time, your state of mind will naturally become more positive because you have improved the relationship with your thoughts.

THE POWER OF THE HUMAN IMAGINATION

Thoughts aren't just what you think in words. They can be images, scenes or even dreams. For example, I want you to imagine a dog eating jellybeans on a surfboard. Or picture yourself in a recent time when you felt confident. These are both thoughts. Though it seems effortless, creating these images in your mind requires complex and highly sophisticated brain coordination. Yet, we often take this extraordinary mental capability for granted. Our brain can generate ideas, concepts and sensations internally, regardless of what's happening around us. This ability is what sets us apart from other species. We can perceive beyond our current situation and envision new possibilities – what is, what has been and what could be. We constantly think in images, and those images hold immense power. What the human imagination has done and continues to do for this world is incredible. Almost everything that we use, see or talk about started in someone's imagination – the chair you sit on, the TV show you watch, our understanding of gravity, even this book you are reading. But imagination is not just about bringing new things into the world. It's also about changing your world. Your perspective. Your mind. As the German philosopher Immanuel Kant put it, imagination bridges our perceptions and our understanding. We do not see the world the way it is; we see it according to our instruments. In other words, we don't see the world as it is; we see it as we are.

A 1962 experiment showcased this brilliantly. Researchers told a group of people that their right arms were being rubbed with poison ivy. Consequently, all the subjects' arms reacted with itching and redness, and some people developed boils. Later, the participants learned that the leaf wasn't poison ivy at all – it was a harmless shrub. The experiment was also reversed. The group were told they were being exposed to a harmless shrub, but in fact their arms were being rubbed with poison ivy. Even though humans are allergic to this plant, only one in six people developed any type of reaction. Just by imagining a different situation, participants were able to influence their mind and body.

There is a fundamental persuasiveness to human imagination, but the problem is so many people do not tap into the power of it, or else they use it in the wrong way. Just like the words you speak, whatever you imagine daily, you are training your mind to become good at. But most of us occupy our mind with worries, stresses and limiting beliefs. We picture the worst-case scenarios or what could go wrong. Just imagine how that is wiring your brain.

Nick, a client, used to vividly imagine freezing up during his presentations or being laughed at in social situations. He would also catastrophise any health symptoms he experienced. All of this fuelled his anxiety daily. Together, we intentionally redirected his imagination. He began visualising positive outcomes – confidently delivering speeches, responding with calmness and enjoying socialising. The more he repeated these images, the more his mindset completely shifted. In four months of training, Nick was able to turn a source of fear into one of his greatest advantages. The power of imagination is astonishing. As Einstein famously said, 'Imagination is more important than knowledge. For knowledge is limited, whereas imagination encircles the world.' Put simply, knowledge is what *is*, but imagination is what *can be* and your brain is the creative engine.

I invite you to start to wonder again. As kids, we used our imagination all the time. It had no limits. But as we get older, we slowly pay less attention to it. It's time to expand your imagination and use this unique and most likely suppressed feature of your brain to level up your mental fitness. The visualisations you will learn and practise in this book will teach you how to be more intentional with the images you are feeding your brain, in turn changing your thoughts from destructive weapons that can harm you to constructive tools that help you.

We can't talk about the mind without having a conversation about emotions. Emotions play a huge role in your happiness, health and performance – they give life and vibrancy to your thoughts, both negative and positive. In the next chapter, we'll look at why it's so important for your mental fitness to not only acknowledge but also fully express and experience all your emotions.

CHAPTER 3

FEEL IT TO HEAL IT

The things we avoid feeling end up becoming the walls we can't move past.

Your thoughts and emotions are inextricably linked, constantly influencing and shaping each other. In fact, an underlying emotion can be the source of hundreds of thoughts. This dynamic relationship not only impacts how you feel in the moment but also affects your mindset in the long term.

From a young age, I was someone who felt emotions very deeply. I still do. Little things, big things – I feel a lot. But I never knew how to properly express them, especially emotions like sadness, anger or jealousy. As I started playing badminton competitively and had insecurities with friendships, I became quite an angry kid, with nowhere to put that anger. Because of the behaviours that had been modelled around me, combined with cultural expectations, I rarely felt safe enough to express my true emotions. They signalled weakness and led me to experience shame. So, naturally, I suppressed them. But I later learned that the body expresses what the mind suppresses. Years of unspoken anger, unacknowledged anxiety, and unresolved resentment were likely to have been key contributors to the inflammation and pain I experienced.

While thoughts are the language of the brain, emotions are the language of the body. Neuroscientist Dr Candace Pert describes emotions as 'energy in motion', which really captures the essence of how feelings move through us. Just take a moment to reflect on that. Many of us try to think away our emotions, but does it ever work? I'm sure we have all been in a scenario where we are feeling stressed or anxious and someone responds by saying, 'Why don't you calm down?' or 'Try to be less anxious'. Even more frustrating, right? Sometimes it even makes it worse! Emotions need to be felt. You want to feel the emotion to heal the emotion. But here's where the real challenge lies.

We have conditioned ourselves to avoid feeling our emotions. When I work with high performers, many of them struggle to understand what it means to 'feel' an emotion. It's because they've become so skilled at pushing emotions aside. It's easier, right? We work them away, drink them away, even sleep them away. The problem is that, when we avoid feeling, the emotions get trapped in the body, sometimes for years, even decades. If you grew up in an environment where emotions were ignored or suppressed, it's likely you've learned to do the same. Or maybe you've witnessed others bottle up their feelings until they explode. Similarly, we've all seen the trope in movies where, when a character gets stressed or sad, they head to a bar for a drink. It's a pattern many of us absorb unknowingly.

But no matter how old you are, how famous you are or how clever you are, emotions do not discriminate. Everyone feels them to some extent. As Tony Robbins says, each of us has an 'emotional home', a place we return to because it's familiar, even if it's negative. For me, that home used to be insecurity and not feeling good enough. Any trigger would send me straight back there. In a way, it became addictive, almost like an old friend. I didn't realise how much it was holding me back. If I hadn't learned to reprogram my mind and use visualisation to better regulate my emotions, I'd probably still be stuck in that cycle of chronic stress and self-doubt.

If you want to start managing your emotions better, these are the three steps to follow:

STEP 1: EMOTIONAL AWARENESS

Emotional awareness starts with being able to detect and identify the emotion you are feeling. It's like a data point. Some people like to label it; others like to just know they are feeling something. I tend to encourage my clients not to label emotions as 'negative' or 'positive' and instead just see them as energy in motion. A big part of awareness is asking yourself, where do I feel the emotion in my body? Do you feel tension in your chest? A knot in your stomach? Maybe a heaviness in your heart? Identifying its physical location is crucial. You don't fear it, you don't amplify it; you just become physically present with it. For those of you who find it uncomfortable or unfamiliar to sit with your emotions, this is going to be hard at first but keep practising. It will help you strengthen your mind–body connection and will be a key foundation to some of the visualisations we'll be exploring in Part 2. Observe your emotions without any intention of doing anything about them.

STEP 2: EMOTIONAL INTELLIGENCE

Once we have awareness, emotional intelligence helps us understand why the feeling has surfaced. Did we get triggered by what someone said? Did we do something that made us feel guilty? Are we fearful of something happening? The longer you ignore or suppress the emotion, the harder it becomes to recall the reason. And it's likely that the more hysterical the emotion is, the more historical the reason is.

For example, an emotion I find particularly hard to feel is jealousy. My body quickly gets consumed, and I feel it in the pit of my stomach. My body experiences a significant physical reaction to it. I never understood why it was so intense until I explored where it came from. Jealousy has been a recurring theme in my life, stemming from my childhood. Growing up with two older sisters, I constantly felt the need to compete for attention. This often manifested as jealousy in my friendships and was heightened even more in the sports arena. Once I realised where this jealousy came from, I was able to extend myself more empathy and compassion. It also allowed me to predict and recognise the situations in which I would feel these emotions.

Sometimes, you won't be able to find a reason and that's also okay! I often meet people who get analysis paralysis at this point – trying to justify every feeling. The point isn't to overthink it; it's just to understand the pattern so we can break it.

One thing I want to add is that a lot of people will skip straight to this step. Have you ever noticed that, instead of letting yourself feel your feelings, you go straight into analysing or intellectualising them? Something upsets you or angers you, but instead of acknowledging it, you go straight to asking what it means or how to solve it. Humans feel like they need to understand something to control it. We have learned to protect ourselves by staying in our head. When we only intellectualise emotions – viewing them through a purely logical lens without actually feeling them – our understanding of these emotions becomes limited. For instance, intellectualising sadness might lead to recognising it as a logical response to loss, but, without feeling it, we miss the deep, internal processing that grief requires for healing. Physical sensations, like the heaviness of sadness, the tightness of anger or the warmth of joy, are part of what helps us understand and release these feelings. That's why steps 1 and 2 are necessary.

STEP 3: EMOTIONAL AGILITY

The final step is emotional agility. This is having the tools and ability to regulate or let go of emotions. This is a crucial skill for leaders, athletes and high performers. Being agile with your emotions is a very personal and experimental process. There are now over 160 distinct strategies we can choose from to change how we feel,* so there is no right or wrong answer here. However, I rely largely on creative visualisation to process and let go of emotions – and you'll come to learn this technique in Part 2. I also fully embrace crying to

---

* Brian Parkinson and Peter Totterdell, 'Classifying Affect–regulation Strategies', *Cognition and Emotion*, Volume 13, 1993, Issue 3, 277–303

release my emotions and change states. In some instances, it serves me better to talk to a friend. Other times, I dance and literally shake the anxiety off. I have had to experiment with different methods to find what works for me:

Sad → connect with others/cry/visualisation
Envy → speak to a friend
Anxiety → creative visualisation or dance it out
High stress → breathwork
Overwhelmed → journal
Anger → creative visualisation

Novak Djokovic, the former world number-one tennis player, explains that when he gets emotionally overwhelmed on or off the court, he accepts it, feels it and comes back. When you watch his games, you will see he is one of the more expressive players in the sport. He sometimes shouts or jumps to release his anger. In turn, he is able to return to a more stable and grounded state instead of letting the emotion build up. Everyone is different.

As we've seen, emotions require flexibility. But you can't do step 3 if you don't do step 1 – it requires a certain level of awareness (even if you don't have a label for it). Let's take a quick look at how this could play out with a common emotion – anger:

1. Emotional awareness: 'I am feeling angry.'
2. Emotional intelligence: 'I am feeling angry because my boss was rude to me and it's made me feel unappreciated.'
3. Emotional agility: 'I am feeling angry because my boss was rude to me, so first I am going to do a short

visualisation to feel and process the anger. Then I will go back and speak to them rationally.'

EXPRESSING YOUR EMOTIONS

When it comes to emotions, it's about expression, not perfection. No one has 'perfect' emotions. A lot of people strive to be happy all the time or think something is wrong if they get sad. Sometimes, we confuse 'being emotionally mature' with not expressing emotions. But that couldn't be further from the truth. Emotions are part of the beauty of being human. Just think about it: we have the extraordinary capability to experience so many different feelings. The more we suppress that, the more we limit the human experience. Granted, they don't all feel amazing, but they definitely make us feel alive.

A useful idea that has helped me in this area is to see emotions as data, not directors. If you wake up feeling miserable, it doesn't mean your whole day has to be miserable. It's simply information about something going on in your life. It doesn't have to direct the rest of your day. But the key with any emotion is to see it as energy in the body that needs to be released or moved. You can only learn to regulate an emotion that you allow yourself to feel. It's not about following a rule book, but it is about learning the language of your mind and body. When I first started working with Sam, a sustainability consultant, he admitted he was emotionally unavailable and that it affected his relationships. We made this a focus for three months, practising awareness, intelligence and most importantly agility. He was able to develop a healthy relationship with his emotions, largely relying on creative visualisation, and in his own words he felt 'emotionally fearless'. This helped him level up his performance and deepen his relationships, both personally and professionally.

At first, it's not that obvious that suppressing our emotions can lead to a decline in confidence, mental performance and ambition, but once you see and, more importantly, experience the connection between the mind and body, it becomes hard to ignore. Visualisation and, in particular, creative visualisation will help you harness the power of your imagination to make your feelings more tangible, giving you a healthy way to process them.

For now, though, let's look at the key piece in the mental fitness puzzle: how limiting beliefs are formed and how you can train your mind to overcome them.

CHAPTER 4

OVERCOME LIMITING BELIEFS

Strength comes not from lifting the weights,
but from lifting the limits you've placed
in your own mind.
– Jessica Ennis-Hill, Olympic and
three-time world champion

I want to introduce you to Rachel. When I first met her, she was deeply affected by the limiting beliefs she was carrying. She believed that in order to succeed in her career as a woman she had to always appear tough. She thought being introverted made her less capable of leading a team and she blamed herself for decisions she had made in the past. These beliefs dictated her actions: she rarely showed vulnerability, avoided leadership opportunities and doubted herself constantly.

Were any of these true? It didn't matter. They were true for her. Through our work together, Rachel began to understand that these beliefs weren't facts – they were stories she had been telling herself for years. She traced them back to comments made by a former boss and unspoken rules in her family growing up. Addressing each belief head-on, with time and visualisation, she was able to let go of these beliefs and instead rewire new ones. Her vulnerability built more trust in her relationships; she embraced leading with 'quiet strength' and got promoted to C-suite level, leading a team of 100. She also now believes that her failures or decisions made her stronger, leaving behind the story of blame she'd held for years.

Our beliefs dictate the way we live our lives, but no one is born with them. We don't come into this world with insecurities, confidence issues or beliefs about money, for example. They are formed from our childhood, family and experiences – whether from our parents and how we observed their actions or perhaps from a comment we internalised from a teacher or boss. Now more than ever, they come from social media – we are constantly internalising different ideas that colour our world.

From a young age, these beliefs become wired in our brain and then dictate how we think, feel and perform. And when these narratives are validated again and again, they create our core beliefs. These are the deep-rooted assumptions we hold about ourselves, the world and others. I recently watched the Disney movie *Inside Out* and its sequel; I loved the way they explained how we develop these belief systems. The film is set inside the brain of Riley, an 11-year-old girl. It personifies her emotions, like sadness, anxiety, joy and fear, demonstrating the complexity of how our thinking patterns and behaviours form. At the start of the film, we see that Riley grew up with strong beliefs about herself, including 'I am a kind person', 'I am loved' and 'I am a good friend.' Her ideas about herself were generally positive as she was raised around good influences and her experiences reaffirmed them. Later, as her external circumstances changed, her core beliefs about who she was, what made her happy and where she belonged were shaken. In particular, she repeatedly had thoughts like 'I'm not good enough' or 'I need to act differently to fit in.' These beliefs soon became a reality to her. The more she bought into them, the more she lost her sense of self and was consumed by anxiety.

Self-deprecating beliefs can create an immense amount of stress and fear in our life. Your RAS will filter and prioritise information that confirms these negative perceptions. For example, if you believe you are not capable or deserving of success, your brain will be more attuned to failures or setbacks, matching your current belief system. This leads to anxiety, low self-worth and declining performance. But this also works the other way. If you believe you are good at something, your brain will notice and highlight all the times you succeed or do well in that area. This boosts your confidence and builds your strengths. Take five minutes here and ask yourself, what are some of your core beliefs that you hold about yourself? Do they serve you or do they sabotage you? And how do they impact your daily life?

Who you are today is a set of programs that was created years ago and probably hasn't changed since. For most of your life, you have left your beliefs to chance or circumstance and probably haven't been aware of them. It's as though you are wearing tinted glasses that colour your world in a specific way. I often notice that my clients find it hard to distinguish between their own beliefs and someone else's. Their lives become subservient to the expectations of others.

For example, someone may pursue a career in medicine or finance because it's what their parents or family expected, even if their true passion lies elsewhere. Or someone told them they wouldn't be able to achieve their goal, so they too believe it would be silly to give it a go. It completely limits what is possible and it's often realised too late. When we are questioned about our beliefs, we tend to respond defensively by saying 'that's just the way I am' or 'I was born that way'. I especially see this among older generations. We've convinced ourselves that we cannot change. But that's not true.

If you don't like your situation, you *can* change it. And if you do like your situation, you can maintain it or make it even better. Your quality of life is a direct reflection of the quality of your beliefs. We all speak about improving our 'mindset', but the word is literally saying that our mind is 'set'. Instead, let's think about it as having a mindflex – a part of us that is dynamic and improving if we train it in the right way. Your beliefs can change; in fact, they need to change. And once you understand that you can choose how your beliefs change through visualisation, you can gain much more control of your life.

If there is one belief that makes the biggest difference to all of us, it's our own self-belief, so let's look now at how to cultivate it.

CULTIVATING SELF-BELIEF

Self-belief is arguably the most important and transformative part of our character. If you have belief in yourself, you have the biggest advantage in life. That's why we have to become so good at it. Self-belief can be both loud and quiet.

It's calm, yet crazy. It takes time to build, but can break in seconds. I see it as a great investment that brings huge returns and rewards.

A few weeks ago, I was on the Tube and a group of football fans came and sat next to me. They were pretty loud and excitable. I quietly observed them. The way they spoke about their team and the players with so much faith, trust and belief made me think, *What if more people did that for themselves?* What if you were your biggest fan? We are much better at believing in other people than ourselves. But why? And what can we do to change that?

My self-belief used to be about 3/10. I relied on external validation and others' opinions to feel capable and confident. Their opinions formed my opinions. My self-belief and self-esteem were also intrinsically linked to my achievements. This was a pattern I internalised from childhood. Only if I achieved the A* or the gold medal would I feel worthy. Today, my self-belief is a solid 9/10 because I have spent the last few years rewiring my brain to fully have faith in my own abilities, despite what others say or the failures I have encountered. It's not a 10/10 yet because I still have some work to do. For example, last week I faced rejection after rejection, and it definitely made me question my self-worth. No matter how many times I hear that rejection is part of the journey, it still feels like a slap in the face. Naturally, it made me feel less motivated and confident about myself. But because I have the tools to build it back up again, I am able to draw on my inner strength and character. It's something I continue to work on every day because self-belief is a skill – a skill every single one of us needs and can improve our lives with. It all comes down to the thoughts you repeat, harnessing your imagination and training your mind.

When it came to cultivating strong self-belief, I first started by becoming aware of and understanding the beliefs I had about myself: the ones that served me and the ones that didn't. This shed a light on what ideas were limiting me personally and professionally. It's been fascinating to observe how my self-belief varies across different situations. While I exude confidence on stage or during networking events, it can waver in other areas of my life, particularly within certain relationships or when confronted with challenging obstacles. The skill was situational, but I wanted to make it more foundational.

To achieve this, I practised visualising for ten minutes every day. This means I visualised and shaped the person I wanted to be, I saw myself achieving my goals, no matter how big or small, and I worked hard to connect to the emotion of what it really felt like to trust myself. From walking again, to building a company, to speaking on stages around the world, I kept showing my mind I could do it even though I wasn't there yet.

I also consciously changed my self-talk to make it more encouraging and motivational. Initially, this felt forced and fake. But I soon grew to enjoy it and feel it.

In turn, I started to believe that I could do anything I put my mind to – whether that was living a healthy life again, hosting my own events or achieving ambitious goals. It felt like this inner knowing and strength. The growth in my character became even more obvious when I was being told by others around me that my goals might not be possible or that they would be hard to achieve. Even when this was happening, I still had my own belief to rely upon, which is the reason I kept going.

Let's go back briefly to that first step: becoming self-aware of your belief system. It's very easy to lie to yourself about what's going on in your mind. The beliefs we hold about ourselves are easy to ignore or even justify. You can't do that with the body; if you are physically injured, you can't hide from that. For one, in most cases you can actually see a physical injury, and so can others. But a belief that you aren't capable or aren't good enough is easy to hide. The problem is that belief will impact your professional and personal lives. But once you are aware of your negative beliefs, you can change them. To do this, carry out a belief audit.

The questions below will help you to identify and reflect on the beliefs you have, why they are there and if you want to make the commitment to change them. Your beliefs are your choice now. (P.S. Be honest with yourself.)

1. Do I believe in myself? Why/why not?
2. What's a belief that got me through a difficult time? Do I still believe it?
3. What's a belief that is stopping me right now? Where has it come from?
4. What is a belief that serves me very well? How can I make it louder?
5. What is a belief I have adopted from my parents that serves me/doesn't serve me?
6. What is a belief I have internalised from social media that isn't helping me?

OVERCOME LIMITING BELIEFS

7. What does it look and feel like to believe in myself? And when did I last feel it?
8. Do I believe I am good enough? Why?
9. What is a belief I have that most people don't have?

I would suggest spending 10–15 minutes journalling on these questions or chatting them through with a friend or coach. Once you have greater awareness about how your brain is wired, the next step is to pick one or two beliefs that you want or need to change. The visualisation practices we'll be exploring in Part 2 will enable you to build new neural pathways to support your new way of thinking and believing. I would particularly suggest outcome visualisations for self-belief (see page 113) and the Batman Effect (see page 117).

One of my favourite stories of overcoming limiting beliefs comes from Roger Bannister, the first person to run a mile in under four minutes. Every single person told him he couldn't do it. In fact, there were scientific research papers documenting how it was physically impossible for a human to run that fast. Even the best runners in the world couldn't achieve it. Then, on 6 May 1954, he ran a mile in 3 minutes and 59 seconds. That led to 50 more people running a mile in under 4 minutes. Isn't that fascinating? Roger had to believe in something that wasn't possible yet – and look at what he achieved. I often think, what does it really take for someone to go against what others say and go for a goal like this? In an interview following the triumph, Roger explained that every single day he would visualise himself achieving his goal. He would even go as far as imagining the time on the clock. It wasn't impossible at all – it was a mental limitation.

Life is too short to let other people tell us what we can and can't do. If you want to create something different for your life, you can make it happen. But you must choose to believe in yourself and new possibilities. Then commit to a belief and practise it. That last part is the most important. I could have written this chapter and told you hundreds of reasons why you should believe in yourself, but that won't make a difference. You need to be the one to wire your own brain and body to believe in yourself, and that takes work and practice. If you are prepared to do the work, then watch your life completely change. Believing in yourself is a full-time job. You get no days off! But it's worth it because it will reward you in so many ways. It creates new opportunities. It builds deeper relationships. It makes you braver and bolder. It gives you the best chance to achieve all the goals you set your mind to. And, more than anything, you are giving yourself the best chance to become the person you want to be.

There are going to be times in your life when no one is in your corner. But if *you* are in your corner, nothing can stop you. You have to back yourself. Think of yourself as your biggest fan. In sport, you have casual fans who turn up to a game or two and generally pay attention when the team is doing well. Then you have the diehard fans who show up regardless of whether their team is winning, losing or getting relegated. Those are the fans who make a difference. You might lose a game now and then, but your spirit doesn't need to break. So don't just become a fan, become a diehard fan of your own life. As I love to tell my clients, get those pom-poms out!

THE NUMBER-ONE REASON YOU AREN'T ACHIEVING YOUR GOALS

Whether you want to become more confident, start a business, win gold at the Olympics, run 5k or just be a bit happier, it doesn't matter what skills you have, how clever you are, how fast you are or who you know – if your mind is not wired for your goal, you aren't giving yourself the best chance to achieve it.

As you've now read, from a young age, we have conditionings that tell us we can't change: we won't make it, we can't do it or we must do something in a certain way. In your day-to-day life, these beliefs directly impact the likelihood of you achieving your goals, which we will delve into shortly.

When it comes to goals, focusing too much on uncontrollable external factors will only set you up for failure. What people don't realise is that your thoughts and beliefs play a much bigger role than you think. This is because you have to grow into your goal. Success rarely comes down to luck or random chances, although arguably it can feel like that. That's why it's good to frame achieving goals as a skill. It's something you can get better at and improve in, and knowing this encourages you to invest in the areas you *can* control – those internal factors, like your character, mental preparation, belief and skillset.

Your inner success cycle

I want to introduce you to your inner success cycle – a framework that significantly improved my approach to goal-setting

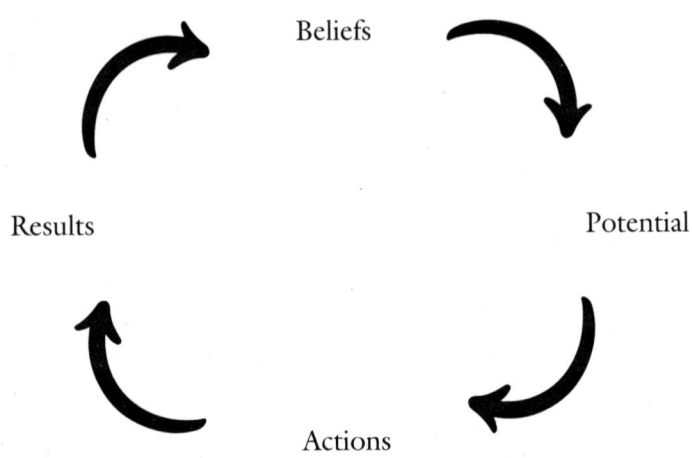

and achievement. Once you understand this cycle, you will approach your goals and beliefs very differently.

The cycle starts with your **beliefs** – how you see yourself and the world around you. It might be useful to frame this aspect as 'certainty' as sometimes the world 'belief' can feel too open. The certainties you hold about yourself directly impact your **potential**, which is your innate ability or capacity to do something. Often this is untapped because of the beliefs we feed ourselves. Your potential directly influences the **actions** you take. With empowering beliefs and expanded potential, you are more likely to take consistent and focused actions towards your goals. This ultimately leads to higher performance and execution. That's when we reach the last part of the cycle, **results**. With aligned and confident action, we get good results. These then reinforce our belief system.

This cycle is happening all the time, with tiny things, but also with bigger aspects like our goals and projects – for example, 'I am an anxious person', 'I'll always be single' or 'I won't be able to achieve my vision.' Because for so many of us it is operating in a way that doesn't serve us, we have to be the ones to take control of it and interrupt it. Remember: you are controlling the narrative every day. For example, let's say you are certain that 'you are a terrible public speaker'. So, when put in a situation where you are asked to speak, you will either say no or perform below your true potential. Perhaps you crumble under the pressure or start the presentation believing people will think you are terrible. This then creates results for you that continue to reaffirm the idea that you are a bad public speaker. But it doesn't have to be the case. By visualising the desired results in advance, you can disrupt the negative patterns and restart the cycle with a positive belief. For example, you can mentally rehearse yourself managing your anxiety and speaking exceptionally on public stages – essentially, the end result of the vision. This will start to build new beliefs, leading to more potential, effective action and positive results.

Most people approach their life and goals operating from history and memory. Very few use their vision and imagination to create new possibilities. Let's now look at how you, too, can be like that minority and put visualisation into practice.

CHAPTER 5

THE POWER OF VISUALISATION

Rewiring your brain isn't just
about fixing what's broken,
it's about unlocking what's possible.

Now you've learned all about thoughts, emotions and beliefs, and how they are interconnected, I want to introduce you to the five different visualisation techniques, all of which have different purposes and outcomes. The majority of people who use visualisation are aware of the outcome and process methods, but rarely do I meet people who know all five (including the best performers and athletes in the world). This is what makes visualisation an incredibly versatile technique for addressing the complexity of mental fitness.

THE POWER OF VISUALISATION

VISUALISATION TECHNIQUE	WHAT IS IT?	WHAT IS IT USED FOR?
Outcome	Creating a clear mental image/rehearsal of the results in advance. It's like seeing the end of the movie.	Increased motivation, rewiring limiting beliefs, increased confidence, character-building, goal-setting and goal-achieving.
Process	Mentally rehearsing a process or task (any size or length) to make it easier, faster or better. For example, public speaking, sports performance or sales calls.	Better performance, increased focus, higher productivity and sharper skills/technique.
Creative	Using visualisation to make an emotion, sensation or feeling more tangible.	Increased emotional awareness, intelligence and agility, managing pain and stress, recovering from injury, and it has been thought to boost health and immunity.
Negative	Mentally rehearsing worst-case scenarios or challenging situations.	Better performance and preparedness, higher resilience and motivation.
Explorative	Generating and evaluating ideas as well as problem-solving.	Efficient problem-solving, generating ideas and boosting creativity.

When I was in hospital and the nurse first asked me to visualise what I would be doing if I wasn't lying there, I didn't realise I was being introduced to an evidence-based mental training technique (not sure she did either). I thought, like a lot of people, that what I was doing was just daydreaming or even wishful thinking. As I mentioned in the introduction, I started by mentally rehearsing walking again. This was process visualisation. I would feel each movement, see myself in a hospital gown, hear the nurses' encouragement and notice the strength in my legs as I regained independence. This evolved into visualising walking outside, then running, adding sensations like shoes on my feet, the sun on my face and the ease of my steps. Alongside this, I practised outcome visualisation, setting my sights on leaving the hospital and going to university. Today it continues to be the reason I stay motivated and inspired by my future. As I discovered more about the technique, I started practising creative visualisation for my day-to-day symptoms and pain. A particular one that helped me was visualising a healthy pink colon every day. Over time, I have incorporated all the different types of visualisations into my week. It remains the cornerstone of my mental fitness, aligning my habits, mindset and future.

EXERCISE – LET'S GET SOUR

I want you to start to experience some of the concepts I have been talking about. After all, we learn the most by doing. What you're about to do is going to open up your visualisation senses and abilities. I want you to observe what happens to your body as you do this exercise:

THE POWER OF VISUALISATION

1. Imagine that you are in your kitchen and you open the fridge. In the fridge is half a lemon. You hold it in your hand. It's cold to touch and you feel the slight toughness of the skin. Notice the texture and weight.
2. Bring the lemon closer to you and take a big inhale. The scent hits you instantly: it's sharp, citrusy and you can even tell how acidic it is going to taste. Now begin squeezing the lemon into your mouth. Each drop lands on your tongue and you taste the intense sourness and tanginess. Keep squeeeeeezing the lemon, until it is empty.
3. Really taste it and let the sharpness linger. Maybe your saliva is increasing? You might feel your cheeks tingling or your face scrunching. Now come back to the present and take a moment to reflect on what happened. Felt very real, right? But here's the thing: you aren't actually holding a lemon, and yet your mind and body were convinced you were.

This is a very quick example of how the mind really struggles to know the difference between what is real and what is imagined. In just 30 seconds of visualisation, similar neural pathways in your brain activated thinking that you were eating a lemon. As you reflect on this exercise, ask yourself which of your senses were stronger. Could you see it very clearly but perhaps not smell it? Could you not see it but taste it intensely? This will give you an indication of how strong your visualisation skills are. If you didn't feel a sensation or weren't able to see the lemon, try once more. Remember – engage all your senses. If you found it hard at

> first, keep going; it's a skill. (P.S. Repeat this exercise with your eyes closed to get an even better effect.)

VISUALISATION VERSUS MEDITATION AND MINDFULNESS

Visualisation and meditation are often used interchangeably, but they are in fact different techniques. I am a huge advocate of meditation and have practised it for nearly a decade. I'd always been aware of it from a young age, but only started properly practising it from the age of 16. While there are some similarities between meditation and visualisation, like having to sit down and close the eyes, their processes and benefits are considerably different. They also have different impacts on the brain. According to Andy Puddicombe, the co-founder of Headspace, meditation is about training your mind for awareness and getting a healthy sense of perspective. It isn't necessarily about becoming a different person, a new person or a better person. It's about learning to observe your mind without judgement. Similarly, mindfulness is the quality humans possess to help them become more present and conscious. This helps to slow the activity in the brain, leading to relaxation and higher self-awareness. On the other hand, with visualisation, instead of trying to clear or slow the mind, you are using your thoughts and imagination for a deliberate outcome – for example, goal-setting, building confidence, improving public speaking and pain relief as well as relaxation and destressing. Returning to the analogy of the mind being a city (see page 30), meditation is a 'deconstructive' technique and can be described as slowing down the traffic, but visualisation is the process of building new roads and breaking down old ones. This makes it a 'constructive' mental training technique.

While meditation and visualisation are distinct practices, they both have an important role to play. For example, when I first started working with one of my clients, she had been practising meditation for a while. So, we designed her morning routine to involve 10 minutes of meditation, followed by 20 minutes of visualisation. We harnessed the benefit of meditation to slow and calm her mind, which meant she could better engage in the visualisation – mentally rehearsing her goals and tasks, getting herself prepared and confident for the day ahead. These techniques and others can all be part of your toolkit; there is no rigid rule book. Bestselling author and coach to billionaires Robin Sharma, for instance, talks about his 'MVP' morning routine. He starts with meditation, followed by visualisation and then ends with prayer. It doesn't have to be either–or. The key is finding what works for you.

VISUALISATION VERSUS MANIFESTATION

In the last few years, I have done over 300 events and keynotes talking about visualisation and, every time, I get asked if it is the same as manifestation. My answer is no. Visualisation has got a questionable rep over the years. We hear stories like Jim Carrey visualising a cheque for $10 million and then that suddenly being his reality. Sounds a bit woo-woo, right? While so much of the story (and hard work) goes missing in translation, it's a big reason why a lot of people dismiss the concept so quickly. Don't get me wrong, I do believe in manifestation – simply because I stand by the fact that if you can see something in your mind, you can make it happen; the mind is where it all begins. But I want to spend time demonstrating how visualisation goes beyond this.

The way a technique is framed directly impacts how you use it, why you use it and the benefit you will get from it. Outcome visualisation in particular is often used as a technique for manifesting because it helps you see what you want more clearly and makes it more prominent in your mind, therefore feeding the RAS system we met on page 28. As your brain rewires, it's no surprise that you may reach your goals quicker or see new opportunities. That's because you are changing as a person. Your actions are changing. The way you perform and show up changes. It's just like building muscle physically – you will be able to lift heavier, run faster, feel better and have a better quality of life. That doesn't mean you aren't working hard or taking action or practising a technique every day. But visualisation extends more broadly with the five techniques. Improving your public speaking skills with process visualisation or releasing anger through creative visualisation is not manifestation – it is deliberate mental training to improve performance and mindset.

As you continue reading this book, I urge you to think of visualisation as a mental training skill, not just a tool for manifesting goals. Framing it as mental training makes it tangible – something you can consistently use in your daily routine to create real results. It's mastering a skill, just like practising a physical movement.

'WHAT IF I CAN'T SEE ANYTHING WHEN I CLOSE MY EYES?'

Aphantasia is the inability to form mental images or pictures of things that are not present. It affects between 3 and 4 per cent of the population. For an aphantasic, it feels like their imagination is blind. For example, if I were to ask you to imagine a beach ball, you would be able to imagine the concept of the beach ball, but would not be able to create the image in your head.

The ability to visualise is a spectrum. You can have high levels of imagery ability, normally meaning you can create shapes, colours and images very vividly, or you can have lower levels of imagery ability.

So, the big question is, if you have aphantasia, can you still visualise? The simple answer is yes. The effectiveness of the visualisation does depend on your mental imagery abilities and experiments have shown that people with higher levels of imagery ability saw more effectiveness in their results. However, the good thing is that you can improve your ability to create mental images. It is a skill that can get better. So, if this is you, I would start with less complex visualisations, then build up, slowly adding more senses and environments. Further, by using photos, memories, drawings and writing, you can prompt your brain before visualising. Lastly, as much as it's about the mental imagery, the feelings and emotions play a significant role too. I work with many people who have aphantasia and have seen effective results.

One of my favourite examples of the power of visualisation comes from Major James Nesmeth. It's widely known that he was a soldier in the US army and loved to play golf. He considered himself an average golfer and generally shot in the low nineties, but was always looking to improve his game. Unfortunately, he was captured in North Vietnam and became a prisoner of war for seven years. During his imprisonment, he lived in a tiny cage, saw no one, spoke to no one and barely did any physical activity. Initially, all he did was pray and hope that he would be released, until he realised he needed to occupy his mind in another way. Nesmeth started mentally rehearsing playing golf every single day. He would visualise the experience to the highest level of detail. He placed himself on the exact golf course he wanted to play. He would smell the grass. He would feel the wind on his skin and the warmth of the sun on his face. He would also see what he was wearing and who he was with. He would feel the club in his hand, with the rough grip on his fingers. He would picture his stance and mentally rehearse taking each swing. He would hear the sound of the ball and watch it fly through the air, landing beautifully. He kept imagining a golf game as if he were physically playing it on the course. He went so far that he even rehearsed getting a drink of water between holes. He did this every day for seven years. He was finally released and came back home to America. Understandably, one of the first things he did was play golf at his favourite course. Even though he hadn't swung an actual club in seven years, he shot a 74! That was 20 shots off his average. For the non-golf fans reading this, that's a pretty massive improvement. Even though he was physically getting weaker and weaker each year, the visualisations he practised strengthened his mental performance and resilience.

This is why the most successful athletes, entrepreneurs and inventors spend time investing in visualisation. Let's dive deeper into their mindset now.

CHAPTER 6

BUILDING A CHAMPION'S MIND

You've got to be the champion before
you become the champion.

In the last few years, I have become fascinated by how champions think. What makes them so resilient? How do they achieve greatness? At first, I believed the champion's mindset had nothing to do with me. After all, every book, podcast and documentary I came across seemed to focus on famous athletes or entrepreneurs. I couldn't see myself in those stories. But then I began to apply their principles – those same tools and habits – to my own life. And something shifted. I noticed real change in my health, performance and confidence. It made me realise that everyone can be a champion, and everyone can develop a champion's mindset. I am a champion because I ran 5k yesterday. I am a champion because I said no to going out last week and decided to rest instead. I am a champion because I am pursuing my dream. I am a champion because I took the leap and decided to go travelling on my own. I am a champion because I make bold decisions instead of being scared. It's not just about being talented or gifted or even winning. It's about being focused, brave and having unshakeable belief in oneself. Your job is to create a vision of yourself and your goals that makes you want to get up in the morning, attack the day and embrace a winning attitude.

WHAT I HAVE LEARNED FROM ELITE ATHLETES

Over the years, I've studied 70 of the world's highest-performing athletes. From Cristiano Ronaldo and Jack Nicklaus to Serena Williams and Michael Phelps – if they're among the best, I've examined their methods. After listening to hours of interviews, analysing their routines and speaking directly to some of their teams, one thing became clear. They all have one secret habit in common: visualisation. Of course, physical practice, dedication and raw skill are vital components of becoming a champion. But the real differentiator that sets the top 1 per cent apart is how they train their minds. Visualisation is the tool they leverage to gain a mental edge. What's interesting is that athletes often don't reveal this secret until they've already reached the pinnacle of their sport. And why would they? It's the secret that gives them an unfair advantage. But don't worry – you won't have to wait that long to learn about it or practise it. I've done the research for you.

I mentioned earlier that visualisation has been most prominent in the sports arena. Soviet athletes were early adopters of visualisation in the twentieth century because they understood how it could maximise their performance capabilities and help them achieve more success. At the Research Institute of Physical Culture in Moscow, the athletes were taught to imagine and mentally rehearse executing different stages of their performance – from their training to winning to working out complex game strategies. Charles Garfield, a renowned psychologist, spent a lot of time with the Soviet researchers. Many of their methods were kept a secret until the release of his book *Peak Performance*, in which he explained that the Soviets' intense training programmes prioritised athletes' mental rather than physical fitness. He cites a study that was conducted to evaluate the success of these government programmes and the results were astonishing. Four groups of elite Soviet athletes were given different combinations of mental training (largely focusing on visualisation) and physical training:

— Group 1: 100 per cent physical training
— Group 2: 75 per cent physical training, 25 per cent mental training
— Group 3: 50 per cent physical training, 50 per cent mental training
— Group 4: 25 per cent physical training, 75 per cent mental training

The four groups were then compared shortly before the Olympic Winter Games in 1980. Group 4 showed the best performance of all the groups. In fact, their improvement was significantly greater than group 3, followed by 2 and then 1.

What did this mean for their success? That year, the Soviets won more medals than any other country – and nearly doubled the total of the USA. While we must also take into account the alleged widespread doping during this period, the proof of visualisation's impact was in their results. From then on, the Soviets continued to integrate psychologists and visualisation specialists into their Olympic teams, making mental rehearsal an essential part of their preparation. This approach sustained their dominance, mental resilience and elite performance for years.

When we look at the greatest champions in sport, business, creativity and entertainment, a common thread emerges that goes far beyond talent, intelligence or even physical effort. It's the fact that they are practising visualisation more than anyone else, leading to the powerful mastery of the mind – whether that was in the 1980s or now. The level of mental fitness, clarity and confidence one can gain by training the mind in this way is unmatched. Let's take a look at some of them in detail.

Tiger Woods, one of the most exceptional golfers in the world, explains in his book *How I Play Golf* that he mentally rehearsed each shot before playing – from the feel of the swing, to the position of his hands, to the sound of the final putt. His visualisation routine enabled him to stay focused under intense pressure as well as improving his confidence and performance. That's what made him exceptional. Visualisation was so crucial to Tiger Woods that, even when he faced significant challenges, such as his car accident in 2021, it remained part of his routine. While bedridden and unable to physically train, he continued mentally rehearsing his golf shots each day. This mental practice allowed him to maintain his skills and prepare for his comeback, showing how visualisation can even aid in recovery and make you stronger.

Michael Phelps is the most decorated Olympian of all time, winning an astonishing 28 medals at the Olympics. He has openly credited much of his success to his mental preparation, particularly visualisation. In a *Forbes* interview, he said his advantage was his mental game and preparation:

> I would visualise how the race could go, how I didn't want the race to go and in a perfect world how I wanted it to go . . . If you are not doing the preparation beforehand, you won't get the results.

Phelps mastered outcome, process and negative visualisation. His coach revealed that he would first start with some relaxation, then he would mentally rehearse. It was like watching a movie. Sometimes he was in the stands watching himself swim. Sometimes he was the one in the water. He would visualise everything from diving into the pool and each stroke to his flip turns and completing the race. By the time he stepped up on the block, he had already swum the race hundreds of times before. His brain switched into what he had visualised. He was always ready for anything that came his way. Any small thing that could go wrong, he was ready for. This is a direct result of his consistent mental training.

In a recent interview with Liverpool.com, Mo Salah explained that visualisation helps him to achieve 90 per cent of his goals. Just think about that: Mo Salah, one of the world's greatest footballers, attributes a huge part of his success to mental training. What I liked most about this interview was his humble admission that sitting down for 15 minutes to train the brain is one of the hardest things to do. Despite being someone who endures serious physical activity, he still regards mental training to be hard. This is a powerful reminder that visualisation doesn't always come naturally – it is a skill that anyone can build and benefit from over time. This goes beyond just sports. Successful entrepreneurs and creators like Sara Blakely, Tony Robbins and Walt Disney are all known to have trained their mind in this way.

Another example I love is Arnold Schwarzenegger. Visualisation is his secret weapon. Arnold didn't just casually imagine himself becoming Mr Olympia; he visualised it daily with relentless intensity. He carried the same mental rigour into every stage of his life. Arnold went from being a bodybuilder to becoming an actor and then a US governor. In his book *Be Useful*, he explains that each time he pursued a new goal or identity, he would make sure to have a very clear vision of what it looked and felt like. Even when he had setbacks (of which there were many), instead of letting them stop him from achieving his goals, he used them as set-ups for the picture in his brain. The vision of who he wanted to become was so engrained in his mind, he was wired to make it happen. I want to stress again: this isn't just daydreaming about something and then it maybe coming to fruition. This is regular mental training. People don't become successful by accident. You have to have that vision and commit to it. In addition to outcome visualisation, Arnold talks about how he used it for his physical training. Before setting foot in the gym, he would visualise himself lifting heavier and better. This combination accelerated his progress.

The athletes we admire aren't just physical powerhouses – they are mental athletes as well. Their success isn't accidental, nor is it just the product of relentless physical training. The secret lies in their ability to visualise, to mentally rehearse their moves, strategies and victories long before they step into the arena.

AN AVERAGE MINDSET...	THE CHAMPION MINDSET...
• makes excuses	• makes it happen
• blames others	• takes ownership
• sees confidence as a feeling	• sees confidence as a skill
• sees failure as final	• sees failure as feedback
• suppresses emotions	• is emotionally fearless
• lets fear control them	• feels fear and does it anyway
• asks, 'How long will it take?'	• says, 'However long it takes'
• uses the phrase 'have to'	• uses the phrase 'get to'
• says, 'It's impossible'	• says, 'I'm possible'
• focuses on being perfect	• focuses on progress
• reacts	• responds
• treats rest as a reward	• treats rest as part of the work
• fakes it	• faces it
• is interested in their goals	• is committed to their goals

Note: The champion mindset doesn't guarantee success. But I can guarantee an average mindset will definitely not get you there. There is an imperfect correlation between optimism and success, but there is a perfect correlation between negative thinking and failure.

Preparation is power

Mental preparation is the competitive advantage no one talks about. When you have a sports match, you physically warm up. Before a big presentation, you rehearse your speech out loud. While all of this is helpful, most people either forget or don't know how to prepare their mind. Have you ever done everything possible to prepare for an important event, whether it's an exam, a match or a presentation, and yet, on the actual day, your performance falls apart? Your mind goes blank and everything you practised seems to disappear. This used to happen to me all the time. When I played badminton, I was great in practice, but, during tournaments, I couldn't match that same level no matter how hard I tried. The night before, doubts would flood my mind, and I realise now that my mental preparation was lacking. I didn't trust myself, and that consistently affected my performance.

Elite athletes are masters at mental preparation. They have specific routines designed to help them feel and perform at their best, especially before high-stakes moments. They also use it to sustain excellence and control how they show up in different situations. For example, the American basketball player LeBron James mentally rehearses his performance just before playing. He visualises different scenarios that might happen during a game, such as how he will defend an opponent or how he will execute a play. By mentally rehearsing these situations, he prepares himself for what could occur during the actual game.

Mental preparation isn't just useful in the context of sports or performance, though – it extends to your relationships, your health, your goals, your emotions and your daily routine. After hearing about how athletes use visualisation to manage opponents or prepare themselves for hard situations, I started using it for difficult scenarios in my personal life. There was one particular situation that stands out. I was breaking up with someone I had been seeing for a while. Often, when we have to have those hard conversations, we end up not saying what we really want or reacting to everything that happens. Instead, I took mental rehearsal very seriously and gave myself the opportunity to prepare my mind for what was to come. I knew the conversation was going to be difficult, so I made sure I rehearsed what I was going to say multiple times. The great thing I noticed was that, when I had to deliver the news and have the conversation, not only was I able to say everything I needed to, but I felt braver and more accepting of the situation.

Pilots have a phrase they use which explains my point exactly. They remind themselves to 'stay ready so you don't have to get ready'. When pressure is high, good pilots rely on their prior preparation and training. As Christopher Voss famously wrote in his book *Never Split the Difference*, 'When the pressure is on, you don't rise to the occasion – you fall to your highest level of preparation.' We all know that preparation doesn't serve you as well when you're cramming at the last minute. It's a much more powerful tool when you make it a regular habit. So, if you want to bring a level of readiness and confidence to your day and tasks, then the mind must be primed and prepared.

This mindset applies to everyday life as well as high-pressure situations. In my own life, as well as preparing for big events and shows, I mentally prepare for my day each morning. Think about it: when you wake up, you brush your teeth, shower and get dressed – you're physically ready for the day. But what do you do to get mentally ready? My guess would be not a lot. Instead, most of us dive straight into social media or emails, getting swept up in the demands of life. Without realising it, we set ourselves up to repeat the same routine day after day, allowing the day to rule us rather than the other way round. The mind becomes susceptible to distractions and brain fog, and we become unproductive, highly reactive and lacking in motivation. The antidote is mentally preparing yourself for the day ahead. It might seem like an insignificant thing, but I've seen a noticeable change since I implemented this habit. I am more focused, productive and relaxed. (For more on this, see 'Mentally Rehearsing Your Day' on page 144.)

Mental preparation isn't just about reaching your peak; it's about staying there, maintaining your performance and managing how you show up in every situation. For example, one of my clients knew she was becoming the CEO of the company she had been working at. Instead of putting off the work until after the change happened, we spent months mentally preparing her with visualisation: placing her in boardroom scenarios, rehearsing key conversations and imagining the work she'd be leading. By the time her promotion came, she wasn't overwhelmed – she was ready. More than that, she thrived in her new role and genuinely enjoyed the experience.

Whether it's tackling a small task or navigating a big life change, mental preparation gives you confidence you didn't even know you needed.

REDEFINING HIGH PERFORMANCE

The phrase 'high performance' is thrown around a lot – in organisations, in the personal development space and in sport. But have you ever stopped to wonder what it actually means? Traditionally, high performance has been synonymous with pushing limits, making sacrifices and relentlessly pursuing goals at any cost. This mindset still dominates many of the organisations I work with, where it's praised and encouraged. But, the more I have understood the essence of high performance, the more I have realised that it's not just about achieving extraordinary results, it is about process, sustainability and balance.

So, what characterises high performance, really? Firstly, high performers know that clarity of purpose is more valuable than speed. If you're running hard in the wrong direction, the distance only amplifies the mistake – that's why knowing where you are going is key. High performance also isn't about saying 'yes' to everything. It's about knowing when to say 'no' and setting firm boundaries. Growing up, I never knew what a boundary was, let alone how to use it. But since I have implemented and communicated my boundaries, I am significantly healthier and happier. I've stopped giving my energy to anyone and everyone. I am ruthless with my time and energy, without feeling guilty for it. I have learned how to monitor my own energy and say 'no' to tasks or people that drain me unnecessarily.

> High performance isn't one-size-fits-all; it is personal and relative. Of course, in contexts like competition, there is a clear scale of performance, but in everyday life, high performance looks different for everyone, shaped by individual goals, rhythms, routines and purposes. At one point in my life, my highest potential and performance was being able to get out of bed and take a few steps. Today, it means building the largest mental fitness company in the world. For you, it might mean managing your emotions better, nurturing creativity or navigating challenges like setbacks, crises or unexpected circumstances. Whatever it means to you, you will be able to respond with more speed and authority when you have mentally prepared and set clear goals and boundaries. That's how you bring out your best performance.

How to avoid burnout: The Seasons Method

One of the first things people sacrifice in the pursuit of high performance or goals is their health. This leads to burnout, fatigue and chronic stress. Today's hustle culture has glorified the idea that working long hours and ignoring personal needs is a badge of honour. I see and hear people boasting about this online, in companies and even among my friends. I used to be the exact same. It was always about doing more and more. It made me feel like a high achiever and performer until it landed me in hospital eight times.

My perspective on this completely changed when I studied how athletes approach high performance. One of the most profound lessons I have learned from athletes about high performance is that you don't have to be 'on' all the time. In fact, that is going to destroy your performance and overall health. Athletes understand the importance of cycles – periods of intense focus followed by intentional recovery. Applying this idea to my life has completely transformed how I work and live.

I call it the **Seasons Method**, and it involves creating three distinct periods in your life: **hibernation**, **growth**, and **harvest**.

During the **hibernation** period, the pace of life is very slow. It's a time to intentionally become less social and more introspective, leaning into a more introverted and reflective state. This is a time when the focus is on planning, consuming ideas and engaging in deep thought. Although overall productivity may appear low on the surface, this is a period of deep work, deep focus and, most importantly, deep rest. This is when you take the time to recover from the efforts of the last harvest season and recharge your energy for what lies ahead. You want to engage in activities that actually give the body and mind good rest and repair – for example, sleep, massages, walks and meditation.

In the **growth** phase, the pace begins to pick up. This is when you would start 'watering the soil' and planting new seeds for the next cycle. It's a time of increased creativity and experimentation, as you begin to implement new ideas and strategies. Your productivity is higher, though not yet at its peak, allowing you to rebuild momentum at a sustainable pace. Social interactions also tend to increase during this time, as you engage and network more with others. Towards the end, you want to rigorously prepare the mind and body for the next harvest season, much like an athlete preparing for a big game. This is a crucial time to practise visualisation because focus, confidence and resilience all need to be strong. The focus here is on refining your energy for the intense work that lies ahead.

The **harvest** season is fast-paced and demanding. This is the time when you would be taking serious action and producing the best results. This will also involve becoming highly social, extroverted and extremely productive. The focus is outward, and you are task-oriented, public-facing and performing at a high level. While burnout is a high risk in this phase, you can avoid it by drawing on the preparation and rest you accumulated during the previous seasons. Further, despite the intensity, you want to make time for little pockets of rest, knowing that balance is key to sustaining high performance.

After the harvest, you intentionally transition back into a hibernation period without guilt. Rest is not a reward for hard work, but a crucial part of the process. I generally repeat this high-performance cycle around 20–30 times a year and adjust the lengths of each season depending on my needs. Sometimes harvest is one day and hibernation is one week. Sometimes harvest is three weeks and hibernation is two days.

Since implementing this cycle three years ago, my performance has skyrocketed, I have not burned out since and I feel stronger than ever. At first, I thought it was just something that worked for me. Then I started sharing it with my clients and the response was really positive. Your seasons might look a little different to mine and that's okay. The lesson here is that by proactively managing your rhythm, you'll reach new levels of performance. The best part is, people in my life are now aware of my seasons and they also respect them.

To start building this into your own life, I would first identify when your next harvest period is. Be specific and note down the days and tasks. Then, immediately after, plan a hibernation period and proactively decide how you are going to rest. Continue with the next seasons from there. A helpful tip is to color-code your calendar and plan out your seasons a few months in advance. Whatever approach you choose, the most important thing is to take control of your rhythm instead of leaving it to chance.

DEVELOPING UNSHAKEABLE CONFIDENCE

Everyone wants to be more confident. I always used to believe you were either born with confidence or not. And I used to think I was in the first category. As a young girl, I was extremely bold, fierce and bubbly. It felt like I had an advantage in life. But with the repetitive hospital visits, inability to socialise or travel and lack of hope for the future, I lost all my confidence. I watched it slip away, little by little, each day, and I had no idea how to stop it. I couldn't make decisions, I didn't know how to socialise, I had no boundaries, I didn't believe I could do certain things and, worst of all, I lost my spark – the thing that made me who I am; my unique fire. I became so unfamiliar with what confidence felt or looked like. It felt very foreign to me – like this thing that other people had, and I could never have again. And that's why I made it my mission to see if I could relearn it. It was time to start from zero.

I spent years studying how to rewire the brain to become more confident. I read every book there was on it. I closely studied confident people – from athletes and champions to performers and CEOs. I explored neuroscience research, looked at psychology experiments and practised for myself. After collating all the information and applying the insights to my own life, I learned that it all comes down to three key pillars:

1. Visualising a confident self-image
2. Mastering self-talk
3. Celebrating little wins along the journey

And the biggest insight that came from everything was that confidence is a skill that everyone can learn, grow and maintain. I've purposely included the word 'maintain' because, just like any skill, you have to keep practising it to stay good at it. Life has a way of testing it. Whether it's harsh criticism, a difficult break-up or a bad performance, it's easy for your confidence to take a hit. Recently, I found myself saying yes to commitments that didn't align with my values, and it left me feeling drained and disconnected. But instead of letting that take control, I relied on the tools I've spent years developing. I visualised the confident version of myself, monitored my self-talk and celebrated every small step towards reclaiming my boundaries. That's how confidence works – you don't just build it once and keep it forever; you must actively sustain it.

So if we break down these three areas a bit more, what do they look like?

Visualising a confident self-image

The first step in building confidence is defining what it means for you. Is it about being assertive and bold? Or perhaps it's about feeling calm and in control. It could be about power and drive. Is it about being able to stand up for yourself? It means something different for everyone and can also change depending on the situation. Ask yourself, what would more confidence allow you to do? What would it really look like? How would you make decisions? Would you do anything differently? According to internationally acclaimed sports psychologist Bob Rotella, champions go out of their way to build a confident self-image. One of the best ways to do this is to practise visualisation. You have to see the person you want to be. Visualising this image again and again helps you to become that person. The visualisation on page 116 will help you to do just that.

Mastering self-talk

The second pillar is one of the most crucial components: self-talk.

Think back to how I described thoughts as weapons in Chapter 2. The thoughts you have about yourself are the ones you can really take ownership of, and this comes into play in the way you speak to yourself too. Your confidence will grow when your internal dialogue is supportive, encouraging and productive.

Meet Priya, a skilled professional who struggled with self-doubt. Despite glowing feedback from her colleagues, her self-talk was her biggest hurdle. At times, she'd find herself thinking, *How could you mess that up? Everyone's going to notice.* Before meetings, she'd tell herself, *You're not prepared enough; someone else could do this better.*

Together, we worked to rewire her self-talk, starting with awareness. Priya tracked her negative thoughts and reframed them:

I made a mistake → *Mistakes are part of learning.*
What if I fail? → *What if I succeed?*
They're ahead of me → *Their path is not mine.*

At first, it felt unnatural, but over time, these shifts became second nature. Within months, Priya stopped spiralling after setbacks and faced challenges with resilience.

Her story proves that by reframing your inner dialogue, you can rebuild confidence one thought at a time.

In Part 3 I will also outline how you can pair different types of self-talk with your visualisations.

Celebrating little wins

Confidence isn't built on massive accomplishments alone. It's nurtured by recognising and celebrating the small wins or victories that happen every day. While I was studying in Bristol, I used to work at a café called Little Victories. Not only did it make the best banana bread ever, but the name also became a reminder for me to focus on the small successes throughout the day – whether it was completing a difficult task, having a meaningful conversation or staying disciplined in my habits. So often we only celebrate huge victories and don't allow ourselves to appreciate the tiny things. These can be as small as 'I ate healthily today' or 'I laughed a lot with a friend'. It could be running for five minutes longer or making a call to a friend you have been putting off. Little victories are endless, and they are an important part of building competence and confidence. I encourage my clients to keep a running list of little victories on their phone. One of my clients now has over 1,000 entries! He explained to me that when he looks back at them, especially if he is feeling a bit low, he finds it hard not to feel confident or grateful because the list reminds him of how far he has come and the progress he has made. Little things can create big changes in perspective. This is something you can start right now. Take out your phone or notepad and start your little victories list.

Confidence versus self-confidence

I want to make a distinction between confidence and self-confidence. Confidence tends to be about external validation – perhaps you walk into a room and someone says, 'Wow, you were so confident speaking', or they tell you how great you look. It increases the more people affirm it to you. This type of confidence is fleeting because it largely depends on others' opinions and situations. Self-confidence, however, is internal. It's about how you feel about yourself, independent of external feedback. Do you trust yourself? Are you secure in who you are? Do you like yourself? When you develop self-confidence, you can walk into a room without needing to be the loudest or most noticeable person.

You feel grounded because your worth isn't tied to what others think. Whether someone likes you or doesn't, it's okay, you like yourself. People make the mistake of trying to 'appear' more confident. That's exactly what I did when I was younger. I appeared bubbly and assured, but internally was feeling extremely insecure.

True self-confidence is an internal game, driven by how your brain is wired. When you invest in yourself internally, the external confidence naturally follows. It's surprisingly easy to erode confidence, which is why being aware of how and when we undermine it is also key to maintaining it. Remember, confidence is just as much about what you *don't* do as it is about what you do. Below is a list of habits, traits and behaviours that either build or break your self-confidence. As you read through it, reflect on how often you engage in each one:

WHAT BUILDS CONFIDENCE	WHAT DESTROYS CONFIDENCE
<ul><li>celebrating little victories</li><li>going on solo dates</li><li>being intentional with your body language (chest out, stand tall, give eye contact)</li><li>helping someone else</li><li>keeping the promises you make to yourself</li><li>setting specific goals and executing them</li><li>creating boundaries and communicating them to others</li><li>being kind</li></ul>	<ul><li>complaining</li><li>gossiping</li><li>compromising on your boundaries</li><li>getting triggered repetitively</li><li>suppressing your emotions</li><li>comparing yourself to others</li><li>shrinking who you really are</li><li>staying around toxic people</li><li>relying solely on external validation</li><li>being chronically stressed</li><li>judging yourself and others</li><li>holding grudges or things from the past</li></ul>

The beauty of confidence is that it's within your reach. No matter where you start, with intentional visualisations, conscious self-talk and the right attitude, you can build unshakeable confidence. It won't always be easy, but the more you practise, the more natural it becomes. And soon, you'll find that confidence is no longer something you chase – it's a part of who you are.

We're nearly at the exciting part – putting visualisation into practice with the techniques outlined in Part 2. Before we put everything we have just learned into practice, let's get clear on what your goals are.

CHAPTER 7

SETTING AND ACHIEVING GOALS

Some people want it to happen, some wish it would happen, others make it happen.
– Michael Jordan

Goal-setting and achievement has always fascinated me – from the psychology of it, to my personal experiences, to understanding how people I look up to have made the impossible possible. There are hundreds of thousands of papers written about pursuing goals and now, everywhere we look online, there is some new advice or framework to achieve them. I'm not here to reinvent the wheel, nor do I intend to bombard you with the latest 'hack' for achieving success. Instead, I want to strip things back and focus on the essence of what makes goal pursuit meaningful and sustainable. Oh, and exciting!

Setting goals and pursuing them is one of the most amazing things we can do as humans. When we bring in too much of the science or practices, we can sometimes lose the inspiration behind goals. I see this a lot when I work with individuals as well as global start-ups and companies. In the pursuit of the goal, they lose the joy of it – the bigger mission they are working towards. This often leads to worse performance and morale. I don't think Einstein was constantly thinking about the SMART goals framework. He simply had a vision, connected to its purpose and then acted. So, if we really strip it back, in its simplest terms, that is the mantra: have the vision and then act – no matter how small. I really mean that. Whether the goal is to run your first 5k or to win the Olympics, the processes in the brain are the same.

As you read this chapter, I want you to start thinking about some of the goals you have already accomplished and some that you want to. Too often, people hesitate to dream big. If they can't believe it's possible, they won't even entertain the idea, let alone pursue it. This is where visualisation has been transformative in my life – and it can be for you too. By mastering techniques like outcome and explorative visualisation, I've accelerated my progress, gained courage in my ambitions, and achieved things I never thought possible.

The mind can either shrink or expand your goals. The choice is yours.

So, when it comes to thinking about your goals, I fundamentally believe it comes down to asking yourself three key questions, which I'll delve into on the following pages.

WHAT DO I WANT?

This might seem obvious, but you'd be surprised how many people don't have an answer to it. In my years of coaching,

I've noticed this happens for two reasons. First, people find it hard to distinguish what they want from what they *think* they want. In other words, are you pursuing your goals or someone else's? This is more common than you might think because of the heavy influence we have from parents, societal expectations and social media. I remember picking certain universities or career paths because I thought they were what I wanted. But when I took the time to re-examine those choices on my own, the answer was very different. I'm quite lucky I figured that out young. Some of my clients only realise later in life that they've chased a goal they never actually wanted to pursue, or that it didn't give them the fulfilment they thought it would. That being said, it's never too late to re-evaluate what you truly want in life.

The second reason is people simply do not give themselves time to ponder the question and therefore don't have enough clarity. When was the last time you asked yourself, what do I want? And no, I don't just mean a new car or more holidays. I mean the thing you actually want to pursue in life. Take a moment now to do exactly this. What's the thing that excites you? Gives you fire in your belly? No matter how old you are or what you are doing in life, this is such a crucial question. And if you don't know the answer just yet, that's okay. When I ask this question at events or keynotes, I often get some people saying they don't know what they want. This is very normal, and it takes time to find the answer. But there are some things you can do to help you find it.

First, start putting yourself in different situations to help you understand what is making you feel alive, inspired or even unhappy. This includes people, work, countries and activities. It will act like a GPS, signalling which direction you want to keep going in. Second, if you don't know what goal to pursue yet, start by pursuing yourself. Work on becoming the healthiest, most healed and most confident version of you. Then, by continuing to ask yourself these questions, you will get the right answers.

Having clarity in your goals is crucial. We live in a world where speed is valued more than direction. But if you don't know where you are going, you will end up coasting or drifting in life. You can have the best ship in the world, but if the captain doesn't know where to go, it will never end up anywhere.

Another thing to be conscious of is the type of goal you are setting. When you set goals that are too easy, it doesn't create enough of a response from the brain or body. On the other hand, if the goal seems too big or lofty (relative to your beliefs), your brain will resist it. The sweet spot is a goal that feels possible yet challenging – something that is just outside your immediate abilities, but still excites and motivates you. It's believed that these 'moderate' goals double the likelihood of engaging in the goal pursuit and achievement. Start with your **aspirational vision**, the big-picture goal. This is the guiding star that shapes everything else. Then, break it down into **long-term goals** (3, 6, 9 or 12 months). These should be ambitious yet realistic stepping stones towards your ultimate vision. Finally, focus on **shorter, immediate goals**, the daily or monthly objectives that keep you on track and make steady progress. The key is layering these goals so that each one builds on the other, creating momentum and alignment across the entire process.

SETTING AND ACHIEVING GOALS

WHY DO I WANT IT?

It's one thing to know what you want, it's another to really know *why*. I know a lot of people who have goals, but no drive or energy or fire for them. Your goals need to be connected to your philosophy and values. Why does your goal excite you? Why is it something you want to dedicate your time and energy to? Why do you want to lose weight? Why do you want to become the top 1 per cent in your field? The why question determines the value of your goal and connects you to your internal motivation. This will help you to stay disciplined, committed and, most importantly, inspired to take continuous action even when things get tough. Challenges are inevitable and motivation will wane, but if you're mentally locked into the reason behind your goal, you'll stay disciplined and inspired, even when the journey gets hard.

In his book *Start With Why*, Simon Sinek explains that we have two parts in the brain: the limbic side, responsible for human behaviours, feelings and value-based decision-making; and the neocortex, responsible for objective or rational thoughts. The aim is to align both these areas of the brain. For example, if you are trying to lose weight, your rational brain knows that going to the gym or eating healthily is the right thing. But the limbic brain focuses on what feels good. It may want to prioritise eating pizza and watching TV on the sofa. When you give voice to the 'why' of your goals, those two systems come into harmony.

So, now that you know what you want, start asking why. When you've asked it once, ask it again. We often get to the core reason after four to six whys. You can do this on your own, with a friend or work with a coach. And if you are struggling to find an answer, reframe the question slightly. What is important to you about the goal? What will it mean for your life if you achieve it or try to pursue it? As you do this, look out for the emotions you feel – the goosebumps, the slight hesitations, the fire in your belly. Our feelings can lead us to our values. And remember, if it makes you happy or inspired, it doesn't have to make sense to anyone else.

WHO DO I NEED TO BE TO ACHIEVE IT?

Goals aren't just about what you achieve; they are about who you become in the process. If you don't believe you're the type of person who can achieve your goal, you won't. It's as simple as that. You will never outperform your hidden self-image. To have the best chance of making it happen, you need to align your identity with the vision you're pursuing.

Further, a big part of visualising my goals is to pinpoint how I want to feel when I achieve them. This strengthens the emotional connection to my goal even more, but it also makes the pursuit more intentional and enjoyable. After all, there's no point achieving something but hating every second of the journey. For example, in my goal of building a global company, I understand it's going to be hard. So, I have decided to approach it as a huge adventure. When there are rejections, I treat it like an adventure; when there are wins, I treat it like an adventure. In contrast, in my goal of writing this book, my chosen intention was to be focused and driven. With the deadlines, research and tasks, my attitude needed to be different. Defining the intention and connecting with the emotions behind your goal will give you strong and meaningful direction.

So, that's really what it comes down to: the what, the why and the who. Of course, there are other things that will make a difference, but I'd argue that most people miss out these three key questions. They're a great place to start and will undoubtedly give you more clarity, connection and competence in pursuing your goals.

LIVE LIKE A PREDATOR, NOT LIKE PREY

In the animal kingdom, there are two types of animals: predators and prey. And you can tell the difference by looking at their eyes. Predators have eyes at the front. They are focused on looking forward and committed to making something happen. Prey have eyes on the side. They are always looking around for danger. They are playing not to lose. Now ask yourself, where are your eyes? They're at the front of your head. You are a born predator, but most of us have been conditioned to be prey. We stop *hunting* for the life we want and start reacting to the life we have.

When I first talk to people about their goals, there's often a sense of hesitation. We don't allow ourselves to dream big anymore or pursue the things we really want to go for. We downplay what we really want, afraid of the disappointment if we don't get there. And I get it – what if it fails? But if you lead with that question, you're already living like prey. Sure, there's risk in chasing what you want, but living in fear is a far greater risk. It keeps you stuck, waiting for someone else to make the first move.

A predator is aggressive with their goals. They don't wait for opportunities to come, they create them. They take ownership of their path, do the hard work and persist even when things don't go according to plan. And here's the best part. A tiger doesn't catch every antelope it goes for, but it is still a tiger. Of course, you are going to miss some goals and face challenges – that's part of the game. Roger Federer, one of the greatest tennis players in the history of the game, gave a speech at Dartmouth College's 2024 graduation day and said one of the best things I have heard in a long time. He quoted that, 'in the 1,526 singles matches I played in my career, I won almost 80 per cent of those matches. But what percentage of the points did I win? Fifty-four per cent.' Even the best of the best lose half the points they play. The difference is, they remain focused on the bigger goal. Federer continued to show up and fight for each point. And, most importantly, he won the points that counted. That is the essence of a predator. This is your reminder to have the bravery and boldness to go for your goals instead of just waiting around for things to happen. In a world full of distractions and pressures, it's easy to lose sight of what matters or settle for a life you didn't truly want. But if you want excellence – whatever that looks like for you – you have to demand it of yourself.

The predator mindset has been the key to my success in many ways. An example that always stands out for me is when I first started Remap Mental Fitness. I knew I wanted to host an event with Nike. I visualised it every day. I knew exactly what it looked like, what the event included and the impact it would have on people. With my target so wired in my brain, I was laser-focused to make it happen. With so much optimism and excitement for the idea, I sent over 15 emails to different people at Nike, but I got nothing back. Undeterred, I sent them again. Nothing. I then started visiting stores in London to pitch my idea. Most people said no or gave me the classic 'I'll send it to the right people', which led nowhere.

Naturally, I felt deflated. But the no's were also making me hungrier. On my fourth visit to a particular store, I decided I wasn't going to leave until I got a yes. Two weeks later, I ran my first event with Nike, and it was a huge success. They asked me why we hadn't done this earlier and then were pitching to me about three more events. This is an example of what the predator mindset can help you do. Because I mentally rehearsed myself achieving the goal every single day, I became laser-focused on the target. I started thinking in new ways, taking bolder actions and became more determined than ever. The inputs and images I was consciously wiring in my brain were determining how I showed up and how committed I was to making it happen. Of course, I wanted to give up at times. I remember the frustrations and low moments. And yes, I doubted myself. But I kept going. Even when I missed some shots or got rejected, I kept my eyes on the goal. Fast-forward two years and I am now Nike's first ever mental fitness trainer. I didn't wait for opportunities to come to me; I hunted them down. At its heart, a predator is someone who just goes for it. They harness their strengths and make it happen. So, start asking yourself, when it comes to pursuing your goals and character, who do you want to be? A predator or prey?

Now you are more aware of how the brain works, the fundamentals of visualisation and what it takes to build a champion mindset, develop confidence and identify your goals, it's time to do something about it. But this is where most people make a mistake. They read books, listen to podcasts, maybe even write some notes, and then they never do anything about it. I am not going to let you make that mistake. Because if nothing changes, nothing changes.

If you really want to take your mental fitness to the next level, the key is to take action and practise. Just like physical fitness, mental fitness is only valuable if you repeatedly practise. If you spent all your life talking about how good running is for the body rather than actually going for a run, you wouldn't get physically fit. Your mental fitness is a daily ritual, not an emergency procedure – even if that means a few minutes a day.

The next section is going to give you everything you need to master the five techniques to think, feel and perform like the top 1 per cent.

PART 2

INTELLIGENCE: MASTERING THE FIVE VISUALISATION TECHNIQUES

Mastery isn't about reaching the end; it's about falling in love with the journey of becoming better every day.

Over the last seven years, I have researched, studied and practised the five visualisation techniques. In that time, I have designed and taught over 1,000 visualisations. One of the reasons I was most excited about writing this book is because there is no single place that outlines each technique, how to do it and what to use it for. In this part, I am going to dive deeper into each visualisation technique and give you specific ways to apply it to your life. As you go through, take the opportunity to practise the visualisations and reflect on your experiences. It's easy to just skip over them, but I strongly encourage you to actually try some – it's so much more valuable than just reading through the explanations. The best way to learn is by doing.

Each visualisation incorporates eight elements, which serve specific functions. This is also why visualisation is different from simply daydreaming or wishful thinking. The practice is specific and intentional, from the words I use, to the way I ask you to prepare the body, to the way it ends.

Let's quickly look at those eight elements in more detail:

1. **Posture:** shoulders and back straight. You want to be in a vertical position of wakefulness.
2. **Close eyes:** removes distractions and allows you to enter your imagination.

3. **Breathwork:** calms the nervous system (remember, neuroplasticity is less effective when you are stressed).
4. **Moment of stillness:** slows activity in the brain and connects you to your mind and body.
5. **Visualisation:** the weightlifting and strength-training part.
6. **Moment of stillness:** helps you to become present.
7. **Breathwork:** regulates the nervous system.
8. **Open eyes:** ready for the day ahead.

By the end of each exercise, you will have taken your mind through a mental workout.

For most of the visualisations, I have added a recommended track that you can use and listen to. These are optional as some people prefer doing visualisations in silence. The music helps the mind and body immerse into the visual imagery. I purposely use quite cinematic and emotional music (without lyrics as this can distract the brain). If you also want to try using your own sounds, feel free to. The music can also create intentional associations that can make your visualisation practice more effective. As you repeat a certain song with a visualisation, the mind can create the imagery or change its state quicker.

I have also suggested the number of minutes needed for each visualisation, but again this can be flexible depending on your own practice and experience. In Chapter 13 we will dive deeper into the optimum times and ways to visualise.

I'm so excited for you to try out these techniques because it's amazing what you can accomplish and how you can feel when your mind is fit. I want to emphasise that these visualisations are not a one-time thing – they require sustained practice and commitment. In the same way you don't get abs overnight, your mind also doesn't rewire itself overnight.

CHAPTER 8

OUTCOME VISUALISATION

A vision is not just a picture of what could be; it is an appeal to our better selves, a call to become something more.

Outcome visualisation is a mental training technique where you are seeing your desired results or the end point in advance. It is commonly used when visualising success, goals or the future. And because the brain struggles to know what's real or imagined, you build a stronger belief that you can do something or become someone. For example, Gabby Thomas, the American runner, won three gold medals at the 2024 Paris Olympics. In an interview, she explained that she had been envisioning winning the races over and over again. So, when she got to the starting block, she already believed she was an Olympic champion. This gave her an incredible level of confidence and resilience – one that clearly allowed her to beat other competitors.

I am no Olympian, but I practise outcome visualisation daily for my goals and have done since leaving hospital. While the goals have changed or been achieved, the technique remains the same. For example, when I was in hospital, I would mentally rehearse being healthy again. I was constantly showing my mind and body what was possible. When I left hospital, I would visualise myself graduating from university. I continue to visualise what my role as Nike's mental fitness trainer looks like and who I need to be to make it happen. I see the end point from now, so I stay aligned to the direction and goal. It gives me motivation, confidence and clarity. My character is also constantly levelling up and getting stronger. But, most importantly, it consistently wires my brain towards success.

HONING YOUR GOALS AND VISION

In Chapter 7, I encouraged you to think about the goals you want, why they are important to you and how you want to feel. Knowing these answers is one thing. Wiring them into your brain is another. This is more important than ever because we live in such a distracted world – life gets busy, people's opinions get louder, doubt loves to creep in, social media makes us dive into comparison holes and, by the time we get to our own goals, we don't have the motivation or energy. That's why athletes spend time visualising their goals daily. They wire their brains for success, laser focus and motivation. If you think back to the success cycle I explained on page 60 (Beliefs → Potential → Actions → Results), this is the visualisation that can help you create a positive and strong success cycle. So, instead of constantly thinking 'What if this goes wrong?' or 'What if I fail?', you can change your thought patterns. By seeing the positive results of your goal in advance, you can build stronger beliefs around it. With these beliefs you will feel your potential and have certainty that you have the capability to do it, which will help you perform better and therefore get better results. This practice is powerful for both short- and long-term goals. Long-term goals might feel distant, but that's precisely why visualisation is crucial — it helps you bridge the gap. When the proof of success doesn't exist yet, your mind is more prone to self-doubt and sabotage. Visualisation fills that gap by creating mental 'evidence' and making the future feel attainable.

The more the visualisation is repeated, the stronger the beliefs become. This is especially important when you haven't physically achieved the goal yet. As you won't have the 'evidence' to support it, you have to train your mind even more, otherwise you will be quick to self-sabotage. The greatest pull on your life needs to be the future instead of being pulled by the past. It's a bit like a magnet – the stronger the goal, the stronger the pull. That's why I am so strict with my clients about making that goal clear. If your goal is to overcome an illness, you want to see what healthy looks like. If your goal is to start a company, you want to be clear on how it looks. If you are an elite tennis player and your goal is to win the US Open, you have to see yourself holding the trophy and celebrating. The vision is only limited by you. Remember, make your goal challenging yet possible!

GOAL VISUALISATION: (5–7 MINUTES)

Recommended track: 'Night' by Ludovico Einaudi

1. **Prepare:** Find a comfortable position and close your eyes. Move your body to release tension, take five deep breaths and a moment of stillness.
2. **Focus:** Start by bringing your attention to the goal/project/mission you are working on. The thing that excites you.
3. **Visualise the end result:** Start seeing yourself successfully achieving the goal. Visualise the results in advance. What does it look like? Where are you? Are you inside? Outside? Is it cold? Hot? See the shapes and colours all around. Place yourself where you will be.
4. **Add details:** What can you hear? What types of conversations are happening? Is there cheering? Is there silence? Who else is there?
5. **Go deeper:** What are you wearing? What are you drinking or eating? Taste the food. What activities are you doing? Remember, this is your goal and you have already achieved it. What does the result look like? Try to see if you get a better picture of what it looks like.
6. **Connect emotionally:** How do you want to feel? Proud? Excited? Peaceful? Fulfilled? Driven? Start bringing this emotion into your body. Combine it with your vision and goal.
7. **Go bigger:** Let's take it up a notch. Imagine going 1 per cent bigger – what would this goal look like

then? What would you do or try if you didn't let fear and doubt take over? Let your mind stretch and explore this bigger, bolder vision. Stay here for a few minutes.
8. **Rehearse an action:** Now, mentally rehearse yourself taking one action towards this goal. No matter how small or big, what is the first step? See yourself taking action. Commit to it.
9. **Close with clarity:** Take five more deep breaths, a moment of stillness and gently open your eyes.

Goal visualisation isn't just for individuals; it's also for companies. Whether you are a small start-up or a multinational corporation, your team can hugely benefit from this. I'd go as far as saying that, if you aren't doing this, you are at a disadvantage. I once heard this story about a CEO of a large corporation in America who sat down with all his VPs. He purchased a 1,000-piece jigsaw and laid out all the pieces. He removed the lid with the final picture on it and asked them to complete the puzzle. After 15 minutes, people were getting frustrated as it was impossible to know where the pieces went. The CEO then said that he would take each person one by one to see the picture in the other room. But he tricked them all and showed each person a different final picture. People began to get even more frustrated and even started arguing. The reason he did this was to show them that, so often, a team can be working together but can't see what the big picture is. I witness this mistake all the time. Without a shared, clear vision, even the most capable teams will falter. When I work with companies, the first thing I ask is, 'Do you have a mission or goal you are working towards?' They tell me yes and then send me a lovely deck or one-pager with their vision statement. That's great and the team clearly have a high-level vision. But does the team actually see it? Or is it just a few words on the company website? Can they envision the result? Does it excite them? Do they believe they can actually make it happen? And, most importantly, do they know their purpose in helping the team get there?

My client Hashim Al-Attas, the CEO of Leylaty Group, has always been an impressive visionary. Earlier this year, I went to Saudi Arabia to work with his executive team to bring this vision to life with visualisation. The team collectively envisioned the company's future and their specific role in it: what success looks like, how their international presence will operate and how they'll uphold their reputation for excellence. This practice not only built momentum and motivation but also deepened the team's connection and alignment.

This exercise is an effective way to bridge the gap between where the team is and where they want to be, with the outcome being that the team immediately feels more motivated and inspired to take action. Doing this for the long term keeps a team aligned and committed. Especially when companies are going through big transformations or structural changes, the goal can seem harder to achieve. But when the vision is compelling enough, your team will be more resilient. You will push through harder moments and not let them defeat you. That's why it's crucial to wire employees for success. There's a difference between having a mission and being on a mission. Teams that are on a mission who can see the end goal with the energy to pursue it will undoubtedly perform better.

Whether you are in an elite sports team, a sales team or a leadership team, you can start adding this to your daily rituals or meetings. For example, start your morning meetings with two minutes of visualisation and watch how it strengthens commitment, boosts morale and sparks a shared sense of purpose. When I have previously suggested this to teams, there is an initial resistance as some people roll their eyes at exercises like this. But all I ask is to try! If the US Navy are doing it, you can too!

INCREASING SELF-BELIEF

As you have already read in Part 1, self-belief is the key to anything we do in life. It's not just about being positive or hopeful about our life and goals – it's about having a resilient trust and certainty in ourselves. It's why my client Anna, who struggled with imposter syndrome, was able to walk into a job interview and confidently showcase her value. It's why Jeremy, a freelance designer, overcame years of self-doubt to raise his prices, attracting higher-paying clients and transforming his business. It's why Leah, a junior athlete often overshadowed by her peers, built the belief that she deserved a spot on the national team and made it happen.

Start to think about where in your life you want more self-belief. Though this visualisation exercise can be used to build general self-belief in lots of areas, I would suggest choosing a specific situation or goal where you are lacking inner conviction and want to increase it. You can also add a few minutes of this after a goal visualisation.

INCREASING SELF-BELIEF VISUALISATION (4–6 MINUTES)

Recommended track: 'My Name Is Barbossa' by Geoff Zanelli

1. **Prepare:** Find a comfortable position and close your eyes. Move your body to release tension, take five deep breaths and a moment of stillness.
2. **Position:** Place yourself in an environment or situation where you normally feel a lack of self-belief. It could be something coming up that feels unfamiliar. It could be a past memory where your mind stopped you from doing something or you felt like an imposter. Immerse yourself in your surroundings. What can you hear? What are you wearing? What can you see?
3. **Frame:** Ask yourself, if I had more self-belief and trust in myself, how would I act in this situation? What would I do? How would I speak? How would I walk? How would I stand up for myself? Watch yourself carrying out these actions in a new way. If you are using a memory, notice the difference in what you would do or how you would act. Show your mind and body what self-belief looks like and create a different or elevated outcome. Go into the details. Get familiar with a new way of showing up.
4. **Add emotion:** How does it feel to have more self-belief? Is it warm? Is it tingly in your chest? Is it strong and exciting? Connect with your body so it

becomes even stronger for the mind. Keep rehearsing the situation until it feels better in your mind and body.
5. **Commit:** Choose to commit to this self-belief. Repeat three sentences of self-talk at the end if you want.
6. **Close with certainty in yourself:** Take five more deep breaths, a moment of stillness and gently open your eyes.

TAPPING INTO UNSTOPPABLE CONFIDENCE

This is by far my most requested visualisation, so I have spent years refining the perfect routine for it. As I mentioned in Chapter 6, I once had unshakable confidence and then I lost it. Rebuilding it was no small feat, but I succeeded by committing to this practice often.

This visualisation goes beyond mindset alone. It works on your body language, communication and energy, training you to embody confidence at every level. If you haven't yet answered the reflective questions on page 88, I'd recommend revisiting them first. These will help you identify what the most confident version of yourself looks and feels like – how they act, dress, speak and carry themselves. This visualisation then bridges the gap between who you are now and the confident person you're becoming. If this sounds abstract, think of it like updating software. Your brain and body won't adopt new patterns unless you deliberately show them what's possible. Many people make the mistake of expecting an overnight transformation, but confidence doesn't work that way. Each visualisation creates subtle, incremental changes that compound over time, building your foundation until confidence becomes a natural part of who you are.

UNSTOPPABLE CONFIDENCE VISUALISATION (5–7 MINUTES)

Recommended track: 'Luminary' by Joel Sunny

1. **Prepare:** Find a comfortable position and close your eyes. Move your body to release tension, take five deep breaths and a moment of stillness.
2. **Build:** Start by seeing the confident version of you in your mind. Build them up from head to toe. What are they wearing? What does their body language show? Chest out, strong eye contact, head high. Look straight into their eyes. See the strength, self-worth and fire.
3. **Create context:** Take this version of you and place them in the situations and environments you will be in. For example, at work, at home, socialising, networking, pitching, playing sport, winning and so on.
4. **Mentally rehearse details:** Start mentally rehearsing how you would think, feel and perform as this confident version of you. What does it look like? How do you approach the activities? What do you say to yourself and others? How do you walk into a room or onto the stage, pitch or court? And how do you feel?
5. **Embrace emotion:** What does it actually feel like to be confident? Energised and fiery? Peaceful and calm? Driven?

6. **Make it specific:** Spend two minutes mentally rehearsing different situations or a particular situation you are focusing on. Add more and more details.
7. **Get present:** Take a moment to bring this confidence to the present moment. Feel it in your body here and now. It might be like a warm buzz or a shock of electricity. Deepen your breath and make this feeling even bigger.
8. **Commit:** Commit to this version of yourself, even if it's just 1 per cent, and bring them into your day/week ahead.
10. **Close with confidence:** Take five more deep breaths, a moment of stillness and gently open your eyes.

The Batman Effect

If you want to take this even further, when I work with individuals, I introduce them to the 'Batman Effect'. The Batman Effect is where you think of yourself as a different character – a person you are striving to be. The concept was initially designed in a study on children, where they were given a persona of a character like Batman to embody when completing a task. The researchers found that children who did this performed 18 per cent better, were 30 per cent more confident and were 15–20 per cent more likely to complete the task. This technique is also commonly used by elite performers. For example, Kobe Bryant used the Black Mamba character because it helped him become an assassin on the court. Adele is said to use Sasha Carter when she goes on stage, which is a combination of Beyoncé's Sasha Fierce and June Carter. It helps her get pumped before shows and allows her to perform better. When we see ourselves as a third-party character, it can be easier to embody confidence, rather than just trying to become a more confident version of ourselves. It's a type of self-distancing.

Research also shows that using third-person self-talk, where you refer to yourself by name or as a character, gives you a new level of control over your thoughts, beliefs and behaviours. By saying, '[Character name] can handle this,' or asking, 'What would [character] do in this situation?' you create emotional distance, reducing self-doubt and helping you level up. I find this to be particularly useful for bigger moments like events or matches as well as big challenges. My own character is called META. I use META mainly for my career aspirations. I even have a picture of her as my phone screensaver. She is a monk and a beast all at the same time. She is bold and fierce, but also extremely kind and joyful. In my visualisation, I watch her walk on stages in front of thousands of people, I feel the warm energy she brings to the room and I watch her smile a lot. I even watch her say no to things and people. She has better boundaries than I do and implements them with confidence. This constantly pushes me to embody this character and become her. In my events, I become the best version of myself and am locked in. It's an incredible feeling. When I start with a new coaching client, this is one of the first exercises I do. No matter what you do, how successful you are or how old you are, everyone can benefit from it.

A few months ago, I met a woman called Andrea who attended one of my confidence workshops. During the session, she explained how she had wanted to leave her job for years, but feared the uncertainty of her future. In the session, we developed her new character, and she continued to do the visualisation for it. Recently, she emailed me to say that she'd finally left her job. She explained that her character gave her the courage and boldness she needed to make a tough decision. One of my other clients played in the 2024 UEFA European Football Championship and we spent a lot of time working on his on-pitch character. He chose a fictional character he has always been inspired by. Before every match, he would spend five minutes listening to the visualisation recording so he could become that version of him. And you could really tell the difference. The way he walked out, the way he played and the way he held himself were significantly better. After the tournament, we reflected on how he felt, and he said it allowed him to enter a zone he had never been able to reach before.

Give it a go yourself by following the steps below:

BATMAN EFFECT VISUALISATION (5 MINUTES)

1. **Choose your character:** This can be a fictional character, someone who inspires you or even an animal. You can also write [your name] 2.0. Be as creative as you would like.
2. **Make a list of your character's attributes:** And remember, if it's too different from you now, it's going to feel unachievable. Be specific. Focus on the beliefs they hold, the energy they bring and the attitude they have.
3. **Prepare:** Find a comfortable position and close your eyes. Move your body to release tension, take five deep breaths and a moment of stillness.

4. **Build:** Start by seeing the character version of you in your mind. What do they look like? What is their body language? Look into their eyes and feel their energy.
5. **Rehearse:** Start mentally rehearsing how your character approaches a task/day/life. Watch them like a movie. How does this version of you speak? Walk? Interact? Exercise?
6. **Add emotion:** Feel the emotion of them. What does it feel like to be them? Fully embody this.
7. **Get present and commit:** Bring the energy, behaviours and mindset of your character to the present moment. Commit to the character. Be excited as you take the steps to become them.
8. **Close with power:** Take five more deep breaths, a moment of stillness and gently open your eyes.

Note: the Batman Effect is not only powerful – it is also super fun! It's a great exercise and visualisation to do with someone else, whether that's a friend or colleague. I would also suggest doing it with a coach as the extra guidance will strengthen the process.

As you keep doing this, you will start to think, feel and perform in alignment with your character. Sometimes, I am faced with a situation and I start doubting myself. I will often ask, 'What would META do? How would she act? What decision would she make?' Because my brain has rehearsed it hundreds of times, I am able to get into that state quickly. I can then make a decision based on a more confident version of me.

When we are faced with challenges, we tend to go back to familiar patterns of thinking and reacting. Aligning to a new character helps you rise up and face tough situations in a new way. My client has a character called Jane. In every difficult meeting, conversation and situation, she embodies Jane. In our sessions, I noticed over time how her language and self-talk started to shift; she would often refer to herself as Jane when discussing her approach. You could see the power this gave her, as she began to adopt Jane's confidence and composure in both her mindset and behaviour. Over time, this approach became her anchor, carrying her through the process with new-found strength.

This nicely brings me on to the next application of outcome visualisation . . .

GENERATING HOPE AND RESILIENCE

Outcome visualisation is particularly powerful for cultivating hope. While hope may sound like a fluffy concept, it is one of the most underrated and integral parts of being human. I have spent considerable time researching hope because it played a pivotal role in my recovery. What seemed like wishful thinking at one point, where I was imagining myself healthy or travelling again, actually turned out to be a specific way of thinking that led me to where I am today.

According to Charles Snyder's theory of hope, it's more than just a feeling; it's a cognitive process made up of three key components:

1. **Goal-oriented thinking:** Knowing what you're working towards.

2. **Pathways thinking:** The ability to find multiple ways to reach your goal.
3. **Agency thinking:** Believing that you can be the one to instigate change.

When you mentally rehearse overcoming a challenge or achieving a goal, you're not just 'hoping' in a passive sense. You're aligning yourself with these three components – giving your mind a roadmap to navigate hardship or a new direction with greater clarity and control.

Life can throw some serious curveballs – illness, business transformation, injury, a failing business, loss, break-ups – or sometimes you can just feel a bit fearful or down about things. In those moments, it can be extremely hard to 'stay positive'. I have found that hope is less about being positive or negative. It's simply about showing yourself that there is a way through. It's not passive emotion; it's an active force. There's a wonderful Victoria Erickson quote I'm often reminded of: 'Hope is the voice that meets you in the storm and says, "There is more than what you see right now."' This type of attitude is life-changing. Many of us rely on external sources for hope – whether it's from loved ones, mentors or even a sense of divine intervention. But developing the ability to generate hope within yourself is a powerful skill. This internal strength can sustain you when external support might be lacking. According to Snyder's research, individuals who scored highly on the hope scale were more successful than others, especially in the sports and academic arena. It's one of the biggest determinants of success and something we can all benefit from.

HOPE AND RESILIENCE VISUALISATION (3–6 MINUTES)

Recommended track: 'Epic Emotional' by AShamaluevMusic

1. **Prepare:** Find a comfortable position and close your eyes. Move your body to release tension, take five deep breaths and a moment of stillness.
2. **Context:** Bring your attention to the current challenge you are facing. Break-up? Redundancy? Feeling stuck? Feel any emotions that come with it.
3. **Cultivate hope:** Start by seeing yourself getting through it and on the other side of the hardship. What does it look like? What are you doing? Who else is there? Rehearse all the different ways you can come out strong on the other side. Create hope in your vision.
4. **Celebrate:** Take it even further and see yourself celebrating after. What would you do? Spend a few minutes here. Feel the relief and gratitude.
5. **Close with hope:** Take five more deep breaths, a moment of stillness and gently open your eyes.

I recently did a keynote for a global financial institution that was going through a big change. New leadership brought unfamiliar dynamics, and overall morale among the team was low. They felt disconnected, frustrated and lacking a clear path forward. I was brought in to help them feel more hopeful about the future. I could have easily got on stage and told a group of 400 bankers to try to be more hopeful in the situation. But that doesn't get results. Instead, I guided them through a visualisation where they actually saw themselves coming out the other side as a team but also individually. I pushed them to go beyond the uncertainty and doubts that were circling round the team and instead see some possible solutions or actions they could take. After the visualisation, the energy was so different in the room. People were more open-minded, they were ready to face the challenges and, more than anything, they were encouraging others around them. Hope gives you a new sense of determination. It is the cornerstone to resilience and is a necessary ingredient for any type of progress. But I am going to tell you what I told them.

Like all the traits I've discussed in this book, hope is a skill. It's something you can cultivate through practice and repetition. Each time you mentally rehearse overcoming your challenges, you deepen the neural pathways of hope in your brain. You train your mind to focus on possibilities rather than limitations. In other words, you build the muscle of hope and eventually become a more 'hopeful' person in all areas of life. And who wouldn't want that?

INCREASING MOTIVATION

One of the biggest misconceptions in society is that you can and need to be motivated all the time. But motivation dies. It's meant to. There are going to be times in your life when you don't feel as motivated. Sometimes we just don't have that strong desire or incentive to do something. And that's okay, it's human. But the question is: do you have the tools to increase it when you need to? You have ownership of your

OUTCOME VISUALISATION

states, and that's the key – knowing that you can dial it up or down. For example, let's say you wake up and want to go to the gym, but you don't feel like it:

- Close your eyes and see yourself at the gym.
- Visualise how you will feel after.
- If you want more of a boost, visualise the bigger fitness goal you are working on – whether that be how you want to look or running a marathon, for example.
- Connect to the goal and remember your why (see page 96).

This will help you tap into your internal motivation. When you open your eyes, you will have a much better chance of actually going to the gym because you have disrupted your normal way of thinking and making excuses. You can use this technique for different situations where you need extra motivation.

    I often work with teams and companies who struggle with motivation. When the workload is overwhelming or there's misalignment within the team, motivation naturally suffers. So, how do you reignite it? How do you motivate your team when they're struggling? And how do you sustain momentum when you're feeling great? By having individuals visualise their personal contributions, as well as getting the team to picture the end goal, clarity and direction are restored. This renewed focus brings a fresh energy to the team dynamic. Many top football teams visualise their ultimate goal for this very reason. Especially when emotion is added, and the team feels more connected, the transformation can be profound. Reflect on how you might use this with your colleagues or your team.

OVERCOMING DOUBT AND FEAR

We often don't take action or do something because we are afraid of what might happen or what others might think. Especially in situations where there is something at risk or you might be going against the norm, it's easy to talk yourself out of it. In fact, we don't just talk ourselves out of it, we visualise the exact situations we fear. By mentally rehearsing negative outcomes, we inadvertently amplify our doubts, feeding our fears even more. Instead, visualising the outcome you want can help the mind to prepare and overcome certain fears or doubts. For example, perhaps you are nervous about going on a date and want to make an excuse so you don't have to go. Or you are thinking of quitting your job, but are worried about finding another one. Maybe you want to go travelling, but are scared to do it alone. One of my friends was performing in a play last week and nearly didn't do it because she was scared about what other people would think. All these fears are normal. But life is too short to let them rule our decisions. Instead, when you are feeling this, close your eyes and visualise yourself getting through the thing you are fearing. See yourself walking out the door and going on the date multiple times. See yourself handing in your resignation. See yourself on stage performing despite what others think, repeatedly. The fears and doubts are simply programs in your brain. You are bigger and smarter than them. It helps you shift focus from fear to possibility, building confidence and making the next step feel less daunting. This doesn't mean fear or doubt disappears entirely, but it no longer holds the same power to stop you.

WHAT TO EXPECT

Immediately after engaging in outcome visualisation, it's common to feel a surge of empowerment, motivation, hope and energy. This technique taps into the deeper layers of your mind, bringing your goals to life in vivid detail. When the vision is deeply meaningful, it's not unusual to experience a strong emotional response, such as tears, joy or a sense of relief. I've cried many times in my visualisations, especially the more real I make them feel. However, don't expect to feel intense emotions every time. Now and then, someone will tell me the effect has 'worn off' because they no longer experience the same feelings they did the first time. The point isn't to chase that intense initial buzz. Your experiences and emotional responses will vary, but that doesn't mean the visualisation isn't having an impact.

It's also natural for people to feel overwhelmed or confused. This can happen if the brain is focusing more on the gap between where you are now and where you want to be. The brain overthinks the 'how'. This can block you from fully embracing the vision. While it's important to have some sense of the path ahead, the focus for these exercises should be on the *outcome*, not every single step (that's process visualisation). A practical approach here is to break your larger goal into smaller, more manageable milestones. For example, if your goal is to run a marathon this year, but you feel overwhelmed by the enormity of it, start by visualising completing 10k or 15k. As you gain confidence in the small goals, you can expand it to a broader vision. If you still find it hard to connect to the vision, I suggest asking yourself if this is really the goal you want.

Every month I host an event in London called Mental Fitness Live. Self-reported data from over 1,000 participants revealed that, after practising outcome visualisation, they experienced a 52 per cent increase in motivation and focus on their goals. Clarity also increased by 47 per cent. These improvements are crucial because they not only increase your chances of success, but also keep you aligned with your true purpose.

In the long term, outcome visualisation rewires the brain towards success, motivation and resilience. Over time, your beliefs about what's possible become deeply ingrained, consciously and subconsciously, reducing self-doubt and redirecting your focus away from fear of failure. However, it's important to remember that new neural pathways take time to strengthen. When you plant the seeds of belief in yourself, don't expect an instant bloom. Like the strongest trees, growth happens slowly, beneath the surface. The bloom is inevitable, but only if you keep watering the seed. That's why repetition is key. Each time you visualise your goal, you reinforce the connection, gradually making your belief system more certain and helpful.

As we explored in Chapter 1, your brain physically changes as a result. One question I always get asked when I teach this is, 'What if I don't achieve my goal?' My answer is simple: it's not only about achieving the goal or outcome, it's about who you become along the way. Life can be unpredictable and sometimes things look different from how we wanted. That doesn't mean the process didn't help.

OUTCOME VISUALISATION

By training your mind in this way and rewiring your beliefs, you will have become stronger, more resilient and energised for life. That's why I often push people to go a little bigger and bolder. When you challenge yourself to envision something that pushes your comfort zone, you stretch your belief system. Whether or not you achieve the exact goal, you develop qualities like tenacity, confidence and creativity.

Lastly, while it's important to see and feel your desired outcome, it's equally important not to *attach* to it. The point isn't that it has to look exactly like your visualisation or be perfect. The point is that it guides you and gives you purpose or direction – something a lot of us have lost. There's immense freedom in pursuing bold, ambitious goals while remaining unattached to the specific outcome. This might seem like a hard pill to swallow, but this balance allows you to embrace the journey, adapt to changes and remain open to unforeseen opportunities.

CHAPTER 9

PROCESS VISUALISATION

I never hit a shot, not even in practice, without having a very sharp, in-focus picture of it in my head.
– Jack Nicklaus, golfer

Process visualisation is a cognitive technique that helps to make any task, process or skill easier, faster and better. It involves mentally rehearsing the steps involved in performing an activity with precision, using all your senses to create a vivid and detailed picture of yourself in action. By doing so, you activate and strengthen the neural pathways involved in the actual execution of the task, which reinforces what's known as a 'positive performance pattern'. Stanford University researcher Karl Pribram called this mental rehearsal the creation of 'mental holograms'. These are 3D images that direct nerve pulses to all the muscles in the body that will be involved in the actual execution of the task.

Lee Pulos, an acclaimed sports psychologist and personal inspiration of mine, used process visualisation when he worked with the Canadian women's national volleyball team. One of the players was working on making a particular shot better. It was good, but she wanted to make it excellent. He guided her to visualise the movement and perfect execution of the shot again and again – from the feel of the ball to the sound it would make when she hit it, to where she wanted it to go, to the position of her body. In a relaxed state, she did this hundreds of times until the mental rehearsal was so vivid and established that, when she started practising the shot on court, her mind and body became one. The result was a nearly flawless execution. She mastered it in her mind and then translated it into physical action.

Process visualisation is useful for improving any skill, from public speaking to performing complex surgeries. And when combined with physical practice, it gives you a substantial edge. The key difference between outcome visualisation (focusing on the goal) and process visualisation is that here, the focus is on *how* you will perform each movement or task. I am going to teach you the exact way to do this.

PERFORMING LIKE THE TOP 1 PER CENT

Process visualisation is the number-one tool for consistent and optimal performance. Sometimes, when we hear the word 'performance', we think it's only for elite athletes or actors. And, generally, a lot of the examples are. That's because we haven't been taught how to apply it to other areas and it's always been a tool the elite have had access to. Performance has so much breadth to it, though. In its simplest form, it is how you do an action or task. We are performing hundreds of these every day and can improve a lot of them. Think about it: you perform when you're in a meeting at work, delivering a presentation, making an important sales call, or even when you're cooking dinner for your family. You perform when you're executing a complex task like a root canal or negotiating a contract. Even going to the gym is a performance! It's all about how well you carry out that action. The reason I love process visualisation is because you can apply it to literally any skill: technical skills, soft skills, hard skills and leadership skills. Mentally rehearsing the action helps you build confidence, improves physical coordination and accuracy, accelerates reaction times and ultimately enhances your craft. As we explored in Chapter 1, your body cooperates with your thoughts and images when you send a clear message, so you can improve your performance without actually engaging in the physical activity.

I want you to think about a task, activity or event you want to perform well in – for example, presenting at a meeting, singing in a show or passing your driving test. Repeat the visualisation below at least five times before doing it.

PERFORMING LIKE THE TOP 1 PER CENT VISUALISATION (3–6 MINUTES)

Recommended track: 'Experience' by Ludovico Einaudi

1. **Prepare:** Find a comfortable position and close your eyes. Move your body to release tension, take five deep breaths and a moment of stillness.

2. **Create the environment:** Place yourself in the environment you will be in when you are performing the chosen task. Are you inside? Outside? What can you see? Who else is there?
3. **Perform:** Mentally rehearse yourself performing the task in the exact way you want to. For example, running, playing the piano, having a conversation, going to the gym, hosting a dinner, serving aces in a tennis game and so on. What does an optimal performance look like? Focus on your movement, pace, accuracy and consistency.
4. **Go deeper:** What part of your body is most intensely involved in performing the activity? What are your hands doing? Are your feet or legs involved? Go as detailed as feeling it in your muscles. Tune into the sensations in your body.
5. **Add details:** What other details are there? Is there anyone watching? Who do you see? Are there other objects involved? For example, a mic? A ball? A slide deck? Hear the sounds and the noises.
6. **Add emotion:** How do you feel performing the activity? In flow? Focused? Energised? It doesn't have to be anything too intense.
7. **1 per cent better:** What does performing 1 per cent better look like? What changes? How do you carry out the action now? Push yourself.
8. **Repeat:** Keep repeating the performance until your mind gets into a flow state with it. See it from the first and third person, repeating the performance the exact way you want it to go.

9. **Close with confidence:** After the mental rehearsal feels established and it feels effortless, take five more deep breaths, a moment of stillness and gently open your eyes.

You can also add physical movements when you do this. For example, there are brilliant videos of Formula One drivers George Russell and Carlos Sainz Jr holding their steering wheels in their hands while visualising. It makes it more real for the mind and body. Last month, I worked with a group of dentists. As I guided the visualisation, some of them held the equipment in their hand. This helped them to find even more clarity in the visualisation.

Using your best performance as a blueprint

Another variation of this visualisation is to recall your best performances. Ask yourself, what are the six to eight actions or things you do when you are performing at your best? For example, if you are a runner, you might think of your fastest or longest run. In that run, what was your form like? What was your pace like? How did your body feel? What was your state of mind? If you are a performer, you might use your best acting scene or show. Again, think about the particular things you did that made it exceptional – your hand gestures, your body state, your interactions. In your visualisation, you want to mentally rehearse that performance again and again, seeing yourself achieve the same standard and result. When you make this your normal, you can level up your overall performance.

Taking this one step further, I want to introduce you to one of my favourite performance tactics that can be combined with the last visualisation. It's called the T-CUP formula, which stands for 'Thinking Correctly Under Pressure', and was created by renowned England rugby coach Sir Clive Woodward. The formula is a combination of three words that acts as both a mental anchor and a trigger, enabling you to access a state of high performance and focus instantly. This is especially useful in high pressure or stressful situations. For example, Michael Phelps uses 'Breathe, Visualise, Execute', Serena Williams uses 'Anchor, Zone In, Attack' and mine is: 'Smile, Visualise, Dominate'. Repeating your own T-CUP formula conditions your mind so strongly that, over time, the mere suggestion of the words to your brain will replay the visualisation, similar to how Pavlov's dogs salivated when the bell rang. This is why you often see tennis players sitting in their chair between games with their eyes closed, repeating certain words. You can prime your brain in those 60 seconds, and it can have an immediate impact on your performance.

How to Create Your T-CUP Formula:

1. **Reflect on what you need most under pressure.** Do you need calm? Precision? Energy? Bravery?
2. **Choose three clear, actionable words.** These should evoke the state you want to embody. Write them somewhere so you remember them!
3. **Practise during visualisation sessions.** Repeat your T-CUP formula as you mentally rehearse achieving your goals.

Apply this same thinking to the way you perform in your personal and professional life. Just before a sales call, use your

T-CUP formula. Just before you go on stage, repeat it. This only works if you have done the visualisation and T-CUP formula enough times for your brain to recognise the association.

OVERCOMING PERFORMANCE ANXIETY

If you are someone who struggles with performance anxiety, I would strongly urge you to make process visualisation a part of your routine. Performance anxiety is a natural and common human experience that can affect many people. Even the best performers, athletes and leaders in the world struggle with it. Often referred to as stage fright, this is a type of nervousness that occurs when you have to perform in any capacity. It might be in front of people, on TV, at a sporting event, reading in class or giving a presentation. It's more than a few nerves; it's when the fear of performing becomes too consuming that it negatively impacts your performance. This type of anxiety tends to come from a previous experience you have either had or observed. It can also be a result of low self-esteem and belief. When the pressure is on, it's all too easy to get worked up. We doubt ourselves, we lose our energy, we rush, we overprepare and we overthink. No matter how good you are at something, if you can't mentally handle the pressure, your performance will suffer.

One of my clients is the Chief Marketing Officer of a global company. She is one of the most competent and confident people I have met. Yet, put her on a podcast in front of a few cameras and her performance suffers majorly. This was harming her reputation and progress. In our sessions, I started by guiding her through visualisations where she was speaking on a podcast or in a public environment. At first the process felt uncomfortable, mirroring the anxiety she experienced in real life. This showed me how deep it ran. It was important not to rush this stage. Instead, we took our time to explore the root cause of her discomfort, gradually helping her feel more neutral in the visualisations.

As we progressed, the next step was to move beyond just neutralising the anxiety to actually improving her performance. We mentally rehearsed podcast episodes over and over. In these visualisations, she focused on maintaining a calm, confident energy while delivering impressive answers with ease. Once she was able to consistently picture herself in this empowered state, we began to introduce positive emotions, such as having fun with the conversation and feeling proud of her performance.

This is a great example of how mental rehearsal can help you unwire certain beliefs, rewire new behaviours and then see excellent results. What had started as a major source of anxiety became an opportunity for her to shine. Now, she's a regular on podcasts, and she can't get enough of them! Think of it like this: she could have pushed herself to go on 50 podcasts and would eventually overcome some of the anxiety. But instead, by mentally rehearsing it 50 times, her mind learned she could actually perform well and feel comfortable in the situation. We made her better at the process in her mind before doing it in reality.

Take a moment to think about a certain situation or task you do in your work or life that brings about performance anxiety. For example, one of my friends recently came to me because she was nervous about a conversation she needed to have with her boss about asking for a raise. Consider how you, too, can use a process visualisation to help you complete the task better.

Improving public speaking

When I was young, I used to be an average public speaker. I wasn't too bad at it, but I definitely didn't enjoy it. I would get consumed by nerves beforehand, didn't know how to handle much pressure and would often choke on my words because of it. Oh, and I can't forget the sweaty palms. Today, I am one of the top public speakers in my field globally, have given two TEDx Talks, spoken on over one hundred stages in front of thousands of people and consistently get exceptional results while having so much fun doing it. I'm not saying this to brag; I'm saying this because visualisation is the reason I was able to improve my skills and standard so significantly. And it continues to be the reason I can keep getting better. If I can do it, so can you. I approach public speaking like an athlete preparing for a big game. My goal is to feel confident, deliver with impact and make my talks memorable. Peak performance requires peak mental preparation. That's why 70 per cent of my preparation involves visualisation.

Let me take you through my exact routine:

1. **Prepare:** I find a comfortable position, close my eyes, move my body to release any tension, take five deep breaths and a moment of stillness.
2. **Create the environment:** I start by creating the environment I will be speaking in. I add the shapes, colours, stage and audience seats. If I don't know what it will look like, I create a general sense of a room.

PROCESS VISUALISATION

3. **Locate myself:** What am I wearing? Where am I standing before walking on? Naturally, I will start to feel a little nervous, but, as I am in a visualisation, I can intentionally train my body into how I want to feel in that moment. I choose an energised calm.
4. **Mentally rehearse walking on:** I then watch myself walking on stage with confidence. I see myself smiling and take a pause before starting. (I used to rush and start as quickly as possible. This signalled nerves. Now, I have trained myself to act differently, and it shows more authority.)
5. **Mentally rehearse the speech:** Then, I mentally rehearse the rest of my performance on stage. I hear the words I will say. I listen to the tone and projection of my voice. I visualise my hands moving as I speak; I continue to see myself moving across the stage smoothly. I mentally rehearse how I want to engage with the audience.
6. **Add emotion:** I ask myself how I want to feel on stage as I perform. I start to embody more confidence and power. I also see myself giving off a warm energy.
7. **Multiple perspectives:** I visualise the experience from my perspective (first-person) and from the audience's point of view (third-person), gaining a full sense of the performance.
8. **Repeat:** I keep visualising the process again and again until it feels very effortless for my mind.
9. **Close with readiness:** I take five more deep breaths, a moment of stillness and gently open my eyes.

Last month, I delivered a keynote to 800 people at a global conference. Starting two weeks before the event, I spent 5 minutes a day mentally rehearsing my performance. Not only does it help me to improve and get better each time, but I am in total flow on stage and my nerves don't consume me. We don't think about the outcome when we are in flow. The mind becomes quiet, tension leaves the body and the performance feels effortless. So many people I speak to either have a fear of public speaking or want to get much better at it. It's a skill that can accelerate your career and life in so many ways. In fact, a lot of C-suites and leaders come to me with this exact problem. They have to speak at big events, but crumble under pressure or don't perform the way they want to. One of my clients is the CEO of a global company. He is often on TV or speaking in front of audiences, but finds it extremely difficult to perform without feeling nervous or stressed, and doesn't have much time to prepare. I have been teaching him this exact process and, last week, I had the chance to watch one of his live speeches. He was sensational. I spoke to him after, and he told me it was the best he had performed on stage and what surprised him most was that he enjoyed it. This was something we worked on in his mental rehearsal. Instead of hating each second, we trained his mind and body to learn how to enjoy it more. Research by cognitive psychologist Sian Beilock, author of *Choke*, supports this. Her studies show that mental rehearsal can mitigate the negative effects of anxiety on performance. By visualising a speech or presentation before the actual event, you create a sense of familiarity for both your brain and body, helping you perform more confidently under pressure.

This technique isn't just for high-profile speakers or performers. I once received an email from someone who attended one of my live workshops on mental fitness. They said they had been using the visualisation recording specifically for speaking up more at work. They would visualise asking questions, standing up for themselves and sharing ideas during meetings. Their email explained that, a few weeks later, so many people had noticed a significant change in their confidence and articulacy in meetings. In another example, I met someone who had a huge fear of public speaking. The thought of going on stage made her feel sick. She made it her goal to visualise herself on stage multiple times, performing the way she wanted to. She practised for ten minutes every day. Four weeks later, she DM'd me telling me she had done five presentations at work and even performed on two stages for spoken word events.

So, wherever you are starting – whether you are at school, regularly pitching to your team or a well-known performer/speaker – try the visualisation a couple of times beforehand and notice the difference it makes.

INCREASING FOCUS AND PRODUCTIVITY

According to Gloria Mark, American psychologist and author of *Attention Span*, human attention spans have gone down from two and a half minutes to forty-seven seconds. Over time, we have trained our minds extremely well to get distracted and lose focus. We then complain that we aren't being productive or getting enough work done. The other day, I was sending some emails and needed a picture from my phone to attach. All of a sudden, 30 minutes had gone by – I was watching some random video on YouTube and had completely forgotten about the emails. Everyone is in the same boat. The sensory and technological distractions all around us are designed to seduce us. And then when we add emotional distractions and mind wandering, it gets even more complex! We find it hard to stay with one thing long enough to really engage with it, sink into it or find that flow. At the same time, small tasks can often feel like huge chores. This is where process visualisation can come into play.

Your attention is one of your greatest assets. And in today's world, a focused fool will achieve more than a distracted genius. When you use process visualisation to improve focus or concentration, it does three key things:

1. Primes the brain.
2. Instigates a flow state.
3. Gives you more cognitive control.

As you visualise a task, your brain begins to treat it as if you've already done it, making the actual task feel more familiar and less daunting. It also helps you access a state of flow, where you become deeply immersed in your work without distraction. This increases your productivity without adding mental strain. Lastly, by mentally rehearsing distractions and refocusing, you train your brain to stay on task more effectively, even when distractions arise.

FOCUS AND PRODUCTIVITY VISUALISATION (30 SECONDS–4 MINUTES, DEPENDING ON THE TASK)

Recommended track: Either no music for total focus or 'Inspirational Piano' by AShamaluevMusic

1. **Prepare:** Find a comfortable position and close your eyes. Move your body to release tension, take five deep breaths and a moment of stillness.
2. **Mentally rehearse the task:** Picture the task you need to complete. It can be as simple as writing and sending a few emails, it can be running an errand, it can be completing a piece of work for your boss or it can be reading for ten minutes. Add all the details. Where are you sitting? Are you on your laptop? What are you wearing?
3. **Visualise completion:** See yourself executing and finishing the task successfully. This is the key part – to repeatedly visualise yourself completing the task until it feels easy and effortless in your mind. Finish the task in your mind again and again.
4. **Spend extra time on challenging tasks:** If you find a particular task difficult or have resistance to it, spend more time visualising that one. Picture yourself staying focused, even if distractions arise, and bring your attention back to the task.
5. **Close with focus:** Take five more deep breaths, a moment of stillness, gently open your eyes and start the task.

MENTALLY REHEARSING YOUR DAY

Visualising your day is an extension of visualising tasks. Just like how you physically get ready, this is the perfect way to mentally prepare yourself. In some ways, it is like mentally rehearsing your to-do list, but with more intention. I also add things like going to the gym or eating healthily in my visualisation. This might seem like a simple practice, but it can significantly help with organisation, energy and having more control of your day. Every morning, I visualise my day before it starts. If I know my day's going to be especially busy, I do this the night before. It makes me more focused and productive.

It's also a good idea to make this slightly more aspirational. Andrew Huberman, the American neuroscientist and host of the podcast *Huberman Lab*, recently interviewed international author and mind–body researcher Martha Beck. In the interview, Beck explained how she spends time imagining her ideal day – rehearsing not only what she needs to do but also how she wants to feel and, most importantly, who she wants to be. Huberman agreed it's an exercise that often goes unnoticed, but is extremely powerful. The brain is a predictive machine, so when you envision your ideal day, it helps guide your actions towards making it a reality. This is especially valuable if you find yourself stuck in a repetitive cycle or feeling uninspired by your daily routine.

The following visualisation can help you become more intentional and excited about your day. The more you do it, the more you will start changing how you carry out your days.

REHEARSING YOUR DAY VISUALISATION (3 MINUTES)

Recommended track: 'I Giorni' by Ludovico Einaudi

1. **Prepare:** Find a comfortable position and close your eyes. Move your body to release tension, take five deep breaths and a moment of stillness.
2. **Visualise the day:** Start mentally rehearsing the string of tasks you need to do from morning to evening, no matter how small, big, boring or exciting (meetings, emails, breaks, cooking, going to the gym, events and so on).
3. **Perform well:** See yourself executing them the way you want to – performing well, maintaining focus and executing with intention.
4. **Add emotion:** How do you want to feel today? Who do you want to show up as? What is your attitude going to be? Start adding this to your visualisation. Show yourself who you want to be. I would suggest choosing things like energised, calm, productive, confident and so on. If at this point you have developed your 'character' from the previous chapter (see page 117), you can bring them in here.
5. **Repeat:** Rehearse the day as many times as you need to until you feel ready to go.
6. **Take action:** Finish by seeing the first action you need to take after opening your eyes.
7. **Close with energy:** Take five more deep breaths, a moment of stillness and gently open your eyes.

When my clients tell me they have back-to-backs and feel overwhelmed, I remind them that they can take back control of their day by doing this in their mind first and then executing. With consistent practice, you'll find that you can focus better, accomplish more and even experience flow in your work. You can also extend this process to visualise longer time periods: your week, month or even an entire season. This ties into the concept of living life in seasons, which I introduced on page 84. For example, I mentally rehearse my entire harvest season at the end of spring so I can start building up more energy, practise my performance and set clear intentions for how I want to move through that time.

ENHANCING SPORTS AND FITNESS

Anyone who is passionate about sports or fitness stands to gain significantly from process visualisation. Whether you're an elite athlete competing on the world stage or someone working to maintain personal health, the process remains the same.

In 2020, Janette Hynes and Zach Turner conducted a study at Transylvania University to explore how positive process visualisation impacts strength training in collegiate athletes. Positive visualisation was defined as mentally rehearsing oneself performing physical movements to the best of one's ability, or even surpassing that – exactly what we've been practising in the previous exercises. The study involved 133 student athletes (70 females, 63 males) from various sports. At the start, the researchers recorded the maximum weight each test subject could lift. The test group was then instructed to visualise themselves performing weightlifting movements, such as the bench press, back squat, clean or deadlift, imagining they were lifting heavier weights than usual while executing each movement with perfect form. The visualisation sessions lasted five to fifteen minutes, were scheduled consistently each day and were accompanied by motivational music. Meanwhile, the control group engaged only in physical training, without the mental rehearsal. After three weeks, both groups continued their regular training routines, and their maximum lifts were measured once more.

The results were striking: athletes who practised process visualisation showed a significant improvement in strength compared to the control group. The test group increased their maximum lifts by an average of 4.5–7kg (10–15lb), while the control group showed a modest increase of about 2.3kg (5lb). Isn't that incredible?

I currently use process visualisation to run further. A year ago, I could barely run for 10 minutes. Now, before each run, I mentally rehearse myself going a bit further. Within three months of consistent practice, I started to run effortlessly for 30 minutes daily. While physical training would've eventually helped, mental rehearsal accelerated my progress because I am neurologically reprogramming myself to perform better.

Entrepreneur and powerlifter Jodie Cook uses process visualisation for her lifting competitions. She visualises every aspect of the day and lift, making the experience so familiar it becomes muscle memory. This prepares her to perform under pressure. I see consistent improvements in athletes at all levels when they apply process visualisation. It helps them outperform themselves and others. And of course, when you are playing on the world stage, this is the difference between being good and being a champion.

Some athletes also use process visualisation after a performance to reflect and learn. After a game or match, they'll close their eyes and mentally review how it went. They analyse what went well, where they could have improved, and what adjustments they could make for the future. This 'mental replay' lets them refine their techniques and mentally prepare for the next competition. It's like giving yourself another chance to play the game in your mind, improving your skills without the physical toll.

IMPROVING SPORTS AND FITNESS VISUALISATION (5–8 MINUTES)

Take a moment to think about an area in fitness or sport that you want to refine or improve. This could be taking a football penalty, lifting heavier weights or a long-term goal such as running a marathon.

Recommended track: 'Pathos' by Ludovico Einaudi

1. **Prepare:** Find a comfortable position and close your eyes. Move your body to release tension, take five deep breaths and a moment of stillness.
2. **Place yourself in the environment:** Start by mentally immersing yourself in the exact environment where you'll perform. Imagine every detail – whether

it's the track, gym, field, court or road. Notice the smells, sounds, temperature and textures around you.
3. **Add more details:** Start to see yourself performing the activity well. Visualise the movements in step-by-step detail: What are your legs doing? How do your arms move? Where is your focus? Picture your body in motion, feeling strong and coordinated. Be as precise as possible; imagine every movement, every adjustment and every position.
4. **Build momentum:** Now, amplify the energy of the performance. Feel the sweat. Get into a rhythm and flow. What does performing 1 per cent or 10 per cent better look like? Rehearse the action again and again until it feels effortless.
5. **Add emotion:** How does it feel in your body? How do you want to feel as you are doing it? Feel the adrenaline. The drive. The strength. See yourself overcoming the harder moments too if needed.
6. **Close with grit:** Once the action and activity feel aligned, take five more deep breaths, a moment of stillness and gently open your eyes.

By following this visualisation regularly, you can start to see real improvements in your performance. The mind and body work together, and with consistent mental rehearsal, you'll begin to achieve better results in less time.

BREAKING AND BUILDING HABITS

Habits, whether good or bad, shape our daily lives. A habit is an action you have repeated enough times for it to become second nature and therefore deeply wired in the brain. By leveraging process visualisation you can reprogram your mind to break unhelpful habits as well as start new ones. Let's start with breaking them.

We all have bad habits. It might be smoking, overeating, biting your nails, procrastinating . . . the list goes on. The key to breaking these habits is disrupting the automatic loop and introducing intentional choice. I'm going to give you a very personal example. When I was recovering from my last flare, at first I didn't leave my house because of the fear of having an accident. Then, for the first few times, I decided to wear a nappy. It acted as a safety blanket, especially in the car or when I went for a short walk. As time went on, I kept wearing them. It wasn't a conscious decision anymore; it was just part of getting ready. A year later, I realised I was still in the habit of wearing them even though I didn't need them. Each day I would visualise myself getting ready and actively throwing away the nappy. I repeated this again and again to disrupt my mind and normal behaviour. After a few weeks, there was one particular day when I stopped myself in my tracks and had the ability to choose differently. After I had done it once, I kept going. This was the new habit I was practising – being able to leave the house without a nappy. I haven't worn them for years.

One of my clients is an international singer. After she performs, she has a massive low. She uses food to comfort herself, but really wanted to get out of the habit of emotional eating. We followed a similar process. As we practised the visualisation, I asked her to see herself picking up the food and, just as she was about to eat it, putting it down. We repeated this visualisation again and again until it felt natural for her to disrupt the automatic emotional eating. Over time, she was able to break the habit. Instead of turning to food, she replaced it with healthier alternatives like writing or connecting with a friend. The visualisation practice helped her take control of her actions in those vulnerable moments.

In addition to breaking bad habits, process visualisation is incredibly powerful for building new ones. When starting a new habit, the mind often resists the change because the action feels unfamiliar. However, by mentally rehearsing the habit, you prime your brain for success.

One well-known thinker, Naval Ravikant, explains how he uses visualisation when adopting new habits. He picks a habit he wants to cultivate, whether it's meditation, exercise or journalling, and visualises himself doing it over and over again. This mental rehearsal makes the habit feel familiar before he even begins, which accelerates the process of building it in real life.

When you see yourself completing the habit in your mind, you're not just imagining it, you're training your brain to adopt it. The neural pathways associated with the habit are forming and strengthening, making it easier to perform and stick to. To solidify habits, combine process visualisation with outcome visualisation. Mentally fast-forward a year into the future and picture yourself having maintained the habit. What does your life look like? What benefits have you gained? This long-term perspective helps you see beyond the immediate struggle and reinforces your commitment. Take a moment now to choose a habit you want to break or start. Spend 3–5 minutes mentally rehearsing the process.

LETTING GO

So far, we've looked at the application of process visualisation to enhance performance. I wanted to include a section on letting go because it, too, is a skill and process. Letting go can be difficult – whether it is a person, a situation, a grudge, a past memory, guilt or resentment. I recently experienced something in my personal life where I felt extremely betrayed and hurt by someone. It left me feeling insecure, ashamed and anxious, completely robbing me of my peace. Like many, I was told, 'You can't control it; just let it go.' As if it were that simple. Over time, I have realised that sheer logic doesn't help me release what I'm holding on to, especially when emotions run deep. What does help is engaging the mind in a process, something tangible yet symbolic to guide it through the act of letting go.

Visualisation can be an extremely effective tool for this. By using imagery tied to the natural flow of water, the brain can process adapting and letting go more easily.

LETTING GO VISUALISATION (3–5 MINUTES)

Recommended track: 'The Tree' by Ludovico Einaudi

1. **Prepare:** Find a comfortable position and close your eyes. Move your body to release tension, take five deep breaths and a moment of stillness.

2. **Locate:** Imagine you are sitting by a flowing river. Listen to the sound of the water, feel the wind in your hair and just focus on how the water is moving.
3. **Mentally write:** Now, in your visualisation, write down the person, situation, event or feeling that you want to let go of on a piece of paper.
4. **Place the paper into the river:** Watch it start to drift away. It goes further and further. Keep watching until you can barely see it. It's so far out, the piece of paper has now gone.
5. **Let go:** Focus back on the flowing water and take five more deep breaths. With each exhale, let go of any emotion or memory that has come up.
6. **Repeat:** Continue this with as many different things as you need to or as many times as you need to.
7. **Close with relief:** Take five more deep breaths, a moment of stillness and gently open your eyes.

If you are sceptical, I understand; I was too. But as I have emphasised, your mind really struggles to know the difference between what is real and what is imagined (see page 25 for a reminder of the science). When you symbolically release something through visualisation, the brain treats it as a real action. Over time, with repetition, the weight of what you're holding on to starts to lift. So, if you find it hard to let go, I urge you to try this exercise. Sometimes, I need to repeat it four or five times to fully feel the release, but, eventually, it works. I can physically, mentally and emotionally let go of whatever I've been holding on to. It's a practice that gets easier the more you do it, and it can offer a sense of relief that is hard to find through logic alone.

WHAT TO EXPECT

Now that we've covered various applications of process visualisation, let's look at the kind of results you can expect. Immediately after process visualisation, people tend to feel more focused, prepared and energised. You may also feel more calm and more in control, but this will largely depend on how you are applying it. If you do process visualisation just before the performance or task, you will notice a change in how you execute it.

For an even greater effect, you can combine process visualisation with outcome visualisation. To do this, first, you see the end result in advance, then you mentally rehearse the processes that are aligned with it. For example, you see yourself crossing the finish line in the marathon, celebrating and building the belief you can do it. Then you mentally rehearse doing the actual run, embodying the movements, the feeling of exertion and the flow of the run. This improves your capability and technique.

In the long term, you can use process visualisation to significantly improve the standard and consistency of your performance, especially with skill-based activities like public speaking and sports. The benefits include improved focus, better resilience and excellent execution. Take Simone Biles for instance. She experienced the 'twisties' (a psychological phenomenon that happens when a gymnast loses their sense of body control when performing or in the air) at the Tokyo Olympics, leaving her fearful to get back on the gym floor or execute any moves. It was so extreme she pulled out of the Olympics. Not only was this heartbreaking for her, but there was also a serious risk she would never be able to perform again at that level. Through intense mental and visualisation training, she was guided to process her routines again and again. She saw herself doing them perfectly and confidently. Four years later, at the Paris Olympics, not only did she execute the hardest vault any gymnast has ever achieved, but she also won three gold medals, something she admitted she couldn't have done without visualisation. This is a great example that no matter where you are in your career or skillset, you can get better at it.

One thing I have noticed when using this method myself and teaching it to others is that, if you are already pretty good at a skill like public speaking or playing a sport, you will see and feel the positive impact on your performance very quickly. Even though the changes may be 1 per cent better, at that level, those small improvements make a big difference. For instance, an athlete might shave seconds off their race time or a speaker might captivate their audience in a new way. Over time, these tiny shifts accumulate into noticeable, significant improvements. If you are using it for a skill you are learning or perhaps aren't as good at, the results may take a little longer to see, simply because you are creating new pathways instead of making existing ones even stronger. With consistent practice, the results come. Process visualisation helps make the unfamiliar feel familiar and the difficult feel achievable.

CHAPTER 10

CREATIVE VISUALISATION

To understand our emotions is to unlock the dialogue between mind and body.

Creative visualisation harnesses the power of the imagination to make a feeling or sensation more tangible. This includes everything from daily emotions and stress to intense physical pain, injury and disease. The goal is to use shapes, colours, characters or objects to help your brain process what you're feeling, or what you'd like to feel, and, ultimately, to release those emotions. This is one of my favourite visualisation techniques, not only for its versatility but also for the immediate impact it can have. Creative visualisation is extremely effective in 40–90-second bursts, making it easy to do whenever and wherever. But when starting to learn it, I would spend a little more time experiencing the process so you can get used to it and strengthen the mind–body connection.

At the majority of events and workshops I do, I guide a creative visualisation. It resonates with a lot of people – from ten-year-olds to sixty-year-olds. This is a technique that encourages playfulness and creativity, allowing each person to take control of their inner world in a way that feels both profound and personal. (P.S. Feel free to be extra creative in these ones.)

MANAGING DIFFICULT EMOTIONS

Humans generally do one of three things when dealing with emotions: we express them, suppress them or try to escape them. High performers or ambitious people often fall into the latter two categories – suppressing or avoiding emotions to stay focused on their goals. But emotions are not something to be ignored. They're part of our human experience and managing them skilfully gives us more control over both mind and body. This, in turn, improves the way we perform, respond to challenges and show up in relationships. Creative visualisation is great for addressing emotional awareness and agility. It encourages you to tune into your body, locate the emotion, feel it and release it (remember, emotions are energy in motion). I practise multiple times a day – when I wake up, on the Tube, before an event or even after a meeting.

Whenever I explain this technique to people, I again use the example of *Inside Out*, as the films personify the major emotions humans feel. They each become little characters. For example, Anger is a short, fiery cartoon man, while Anxiety is this slightly hectic and loud thing. In a similar way, when visualising an emotion, you can make it into a character – for example, an angry red goblin or a sad blue animal. But you can also stick to something simpler like a purple circle or fireworks. There is no limit to your imagination. This is also a great visualisation to introduce to young kids. It's a playful yet powerful way to help them understand and work on their emotions. I wish I'd known about it when I was younger as it would have given me a safe way to feel what I was feeling.

I've chosen to use anxiety as an example here, as studies show that over 40 per cent of people worldwide report feeling anxious regularly. However, the process I outline can be directly applied to other emotions.

Releasing anxiety

Before jumping into how to do the visualisation, I want to discuss the difference between being anxious and having anxiety. Being anxious is something every human will experience from time to time. It's a temporary feeling and often is a normal response to an abnormal or unfamiliar situation – perhaps you are about to go on stage or have a difficult conversation. According to Dave Alred, author of *The Pressure Principle*, this is a type of 'state anxiety', where it's very dependent on the situation. In contrast, having anxiety is more of a permanent feeling. It's when you get anxious feelings in situations you normally wouldn't. It starts to consume your everyday life. This is known as 'trait anxiety'. The good thing is, creative visualisation can be used for both. But, for the latter, I would recommend coupling this with outcome visualisation and therapies that address root causes. In that way, you really get to the core of the beliefs.

RELEASING ANXIETY VISUALISATION (4–6 MINUTES)

1. **Prepare:** Find a comfortable position and close your eyes. Move your body to release tension, take five deep breaths and a moment of stillness.
2. **Locate the anxiety in your body:** Take a moment to think about what is making you anxious. Bring it to the present. Notice where in your body you are feeling the anxiety (remember, emotions are the language of the body). Is it in your head? Gut? Chest? What does it feel like? Is the sensation heavy? Tight?
3. **Visualise the anxiety:** Now, visualise what this anxiety looks like in your body. Give it a colour, a shape or even make it a character/animal. Is it hot? Is it cold? It could be a green goblin on your shoulder, an ice cube in your head or a fire in your belly, for example. Embrace it – this is your imagination's way of processing the emotion. And now just sit with it. Don't try to run away or ignore it. Let the feeling do its thing.
4. **Let it get bigger:** This might get a bit uncomfortable, but stay with it. Allow the shape/character to get bigger and bigger. Feel it.
5. **Shrink the image:** And now, using your mind, let this image get smaller and smaller. You can be as creative as you like with how you want to make this happen. Do you pour water on it? Do you kill it? Do you just let it dissolve?
6. **Repeat:** Keep repeating this until the image gets so small it's the size of a pea.

7. **Release:** Take a deep breath and, on the exhale, let the image fully dissolve and be released.
8. **Locate again:** Take your awareness back to the area in your body and notice the change you feel.
9. **Repeat:** Continue this as many times as you need to until the intensity of the anxiety becomes less consuming, and you feel relief.
10. **Close with lightness:** Take five more deep breaths, a moment of stillness and gently open your eyes.

This creative visualisation helps transform invisible emotions into tangible objects you can interact with and, ultimately, control. What I love most about it is how quickly it works, especially in moments of crisis. It's simple, effective and gives you a sense of control while fostering a healthy acceptance of your emotions. I've seen it help clients with everything from anger and sadness to jealousy, frustration and fear.

One client I worked with had been carrying guilt for years. She had buried it deep within herself, unable to process it fully. Through visualisation, she imagined the guilt as a sharp, spiky yellow ball, a representation of the weight and discomfort it caused her. We then visualised smoothing the ball with sandpaper, gently grinding away at its roughness, and slowly shrinking it down to a manageable size. As she mentally worked through this process, she felt the guilt becoming less overwhelming and more something she could hold and control. Eventually, that massive weight she had been carrying was lifted.

The power of this method lies in its ability to help you coexist with your emotions, rather than suppress them. Over time, this visualisation helped my client release the guilt, and the emotional charge behind it dissipated. This isn't about getting rid of emotions like guilt or anxiety. It's about learning to locate them, feel them fully and release them when you're ready.

FEELING GRATITUDE AND JOY

Creative visualisation isn't just for processing challenging emotions; it can also help you experience and extend the positive feelings you want to cultivate, like peace and joy. We tend to hold on to difficult emotions but often bypass the positive ones. I'm guilty of this too. I'm quick to move on. Personally, this visualisation has helped me relearn how to feel emotions like joy and gratitude. After facing illness and trauma, I became numb to life's happy moments. I couldn't remember what joy or love felt like and how to fully appreciate them when they appeared. Now I experience them to their full capacity and it's incredible.

One emotion I focus on is gratitude. Even as we chase bigger goals, it's important to pause and appreciate where we are. It's easy to forget this. I like to add a creative gratitude visualisation after my outcome visualisations each morning.

GRATITUDE VISUALISATION (3–5 MINUTES)

Recommended track: 'Life' by Ludovico Einaudi and Daniel Hope

1. **Prepare:** Find a comfortable position and close your eyes. Move your body to release tension, take five deep breaths and a moment of stillness.

2. **Locate the emotion:** Start by thinking about something you are grateful for. Notice where in your body you are feeling the positive emotion (remember, emotions are the language of the body). Is it in your head? Gut? Chest?
3. **Visualise the gratitude:** Now, visualise what this feeling looks like in your body. Give it a colour, a shape or even make it a character/animal. Is it hot? Is it cold?
4. **Expand it:** Now allow the shape/colour to get bigger and bigger. Let it expand and spread.
5. **Optional:** Add another situation/person you are grateful for. Repeat the same process.
6. **Enjoy:** Just sit in this for a few moments and really enjoy it. Then bring it back to the one point you started with in your body.
7. **Close with gratitude:** Take five more deep breaths, a moment of stillness and gently open your eyes.

CONNECTING TO YOUR INNER CONFIDENCE

I recently hosted a confidence workshop for 25 women. After doing the Batman Effect visualisation with them (see page 119), I trialled creative visualisation for confidence. I had done this myself before, but not with others. The response was powerful. Following the same structure as the exercise above, the women visualised what confidence looked like in their body. The images ranged from sun and fireworks to lions and stars. As we expanded the image, the feeling became more prominent. I was particularly touched by a woman who told me that she couldn't remember what being confident felt like so she found outcome visualisations hard. That session was the first time she connected with it again. I encouraged her to keep repeating this exercise so she could get familiar with the feeling. From there, she could add things like the Batman Effect. A few months later, she sent me an email explaining the positive change she and her husband had noticed in herself.

If you want to try connecting to your confidence in a more creative way, try this for yourself. Aim for about three to five minutes and really enjoy the feeling it creates in your body. Let it expand. It's great to do this visualisation when you are on your own and want to gain inner power and strength. It's also great to add on to the character visualisations.

MANAGING PAIN

I first started using creative visualisation for the pain in my stomach and colon. It paralysed my whole body and just felt inescapable. Even medication like morphine barely gave me any relief. Naturally, I was very afraid of the pain and going through it every day for seven years really broke me.

I started using visualisation to help me. I closed my eyes and started imagining what the pain looked like in my body. To me, it was hundreds of hungry piranhas eating the inside of my stomach. And they wouldn't stop. So, in my visualisation, I would fight off the fish and shoot them dead. I know it sounds a bit weird, but again, my brain really struggled to know if this was happening or not. When I did this for the first time, I felt a tiny bit of relief. As I continued to repeat the visualisation and process, slowly my pain started getting a bit better day by day. Whether it was actually getting better I do not know, but my ability to handle it and mentally stay strong through it definitely was – and that was enough for me. I relied on this tool nearly every day during that time. Even though I haven't experienced this type of excruciating pain for at least five years now, I still use it if I ever get injured or am on my period, for example.

Because the mind and body are deeply connected, many believe creative visualisation can offer relief from pain. However, when the pain is intense, always seek medical advice. Visualisations can help hugely, but don't be too hard on yourself if the pain makes it too hard to focus on visualising, as sometimes survival is all we can think about. My friend Alex struggled to use visualisation during his chronic illness flare-ups. Instead, he practised it during calmer periods, using it as a tool to better manage the pain when it inevitably resurfaced.

RELIEVING PAIN VISUALISATION (5–10 MINUTES)

Recommended track: Either no music or 'Buddha's Flute' by Buddha's Lounge

1. **Prepare:** Find a comfortable position and close your eyes. Move your body to release tension, take five deep breaths and a moment of stillness.
2. **Locate the pain:** Start by noticing where in your body you are feeling pain or discomfort. Is it sharp? Dull? Throbbing? Focus on where it's most intense.

3. **Visualise the pain:** If the pain was an object, shape, animal or colour, what would it look like? Is it hot? Is it cold?
4. **Dissolve the pain:** With each breath, imagine the pain slowly shrinking, softening or dissolving. You might see the shape change, or perhaps there's a more aggressive intervention. Be creative here. Make it lighter.
5. **Release the pain:** Visualise the object/shape fully dissolving and disappearing.
6. **Repeat:** Keep repeating this until you feel some relief.
7. **Close with relief:** Take five more deep breaths, a moment of stillness and gently open your eyes.

RELIEVING STRESS

Stress is often cited as a key contributor to many serious health conditions, and for good reason. Much like emotions, most people try to think their stress away or ignore it, but that rarely works. When we experience stress, our sympathetic nervous system is triggered. This system governs our fight, flight or freeze responses, preparing us to face a perceived threat. As a result, cortisol levels rise, mental performance drops, fatigue sets in, and, over time, burnout follows. In states of overwhelm, anxiety or worry, many struggle to know how to respond. Instead, they ignore it and push forward. The problem is, even if you forget about the stress, your nervous system doesn't. High performers in particular tend to thrive on stress without recognising its long-term consequences. We get caught in the cycle of working hard, taking a holiday, getting sick during that break, and then never being able to actually destress.

I once worked with a client who presented herself exceptionally well, always appearing so put together. But as we spoke and I learned about her habits, it became clear that her nervous system was perpetually in overdrive. She was in a constant survival mode. The issue is, when everyone around you seems to be functioning this way too, you start to accept it as normal. But it's not.

Don't get me wrong: a certain amount of stress is necessary and even helpful for survival. However, stress becomes dangerous when it's ignored or allowed to build up unchecked in the body. The key to managing stress isn't about wishing it away or waiting for life to get easier; it's about learning how to equip yourself to handle it more effectively.

When it comes to using visualisation to relieve stress, there are a few effective approaches. One method is similar to the Releasing Anxiety visualisation (see page 159). You identify the stress in your body, attach a shape, colour or image to it, and gradually shrink it. Another option involves mentally transporting yourself to a peaceful environment, allowing time to reset both your body and nervous system and truly relax.

RELAXATION VISUALISATION (5–30 MINUTES)

Recommended track: 'Buddha's Flute' by Buddha's Lounge

1. **Prepare:** Find a comfortable position and close your eyes. Move your body to release tension and take five deep breaths. With every exhale, release any tension

or tightness in your body. Unclench your jaw. Soften your eyes. Then take a moment of stillness.
2. **Locate:** Imagine that you are sitting on a bench by a lake. Feel your feet on the ground. Feel the sun on your face. It's warm. Feel the gentle breeze on your body. Listen to the birds overhead.
3. **Awareness:** Now, focus on the water in front of you. It's extremely blue and calm. You can hear the gentle flow of it. There is no one else around, there is nowhere else to be and nothing else to do. Just be.
4. **Stillness:** Stay here for a few moments until you find complete stillness in the mind and relief in the body. If your mind starts to wander, bring it back to the lake and your feet on the ground.
5. **Immerse in water:** Now, imagine the water filling your whole body. Start from your head. It's cool and relaxing. It flows from your head, to your chest, your arms, your stomach, your legs and your feet. The water relaxes your body.
6. **Close with calmness:** Stay here for as long as you like and, when you are ready, take five more deep breaths, a moment of stillness and gently open your eyes.

To take this even further, you can imagine a blue light from the top of your head moving all the way down to your feet. The light enters every cell, every organ, every muscle. The blue light is slightly cool in temperature, so it relaxes your muscles and revitalises all your cells. I do this with my clients before each session.

The reason this can be so effective is because, regardless of the environment you are in, for example a stressful meeting at work, a cramped train or a busy day, the visualisation places your body elsewhere to take you out of survival mode. Once you open your eyes, you can approach the same situation but from a calmer place.

A common challenge I hear is, 'I don't have time for this.' I understand – when we're stressed, stopping for even two minutes can feel like too much. But if we don't take these moments, the stress accumulates in our bodies. It's crucial to take action sooner rather than later. The more you practise this, the easier it becomes to shift from survival mode to a calmer, more centred state. Think of it like recharging a battery. Most of us wait until we're at 5 per cent before recharging, at which point we're crumbling under the effects of stress. But what if you recharged at 40 per cent or even 60 per cent? You'd stay ahead of your stress, proactively caring for yourself.

I have found this exercise to be particularly useful for sleeping. Insomnia is on the rise, especially among young people. Often it's a symptom of underlying stress and overstimulation. Studies have shown that when you relax the body through visualisation before you sleep, you wake up less, enter a deeper sleep and feel more rested in the morning. Who wouldn't want that?

WHAT TO EXPECT

Creative visualisation tends to have a very immediate effect, especially when used for emotional management, relaxation or pain relief. You will feel lighter, calmer, more comfortable and open. For example, when I am guiding individuals for pain relief, they feel a change immediately. Some people even feel emotional or happier as you are literally releasing energy from the body. It can be a big relief. When I carry out these exercises in my events, people are always so surprised at how different they feel. But the best part is hearing how imaginative everyone's visualisations are. We have had green goblins coming out of urine, legs made of steel being broken with a hammer, bugs crawling over the skin and then being flicked away, little soldiers killing cancer cells . . . the list goes on. It just shows how unique each of us is. And with creative visualisation you want to really embrace this.

In the long term, creative visualisation helps you develop a better relationship with your mind and body, especially when it comes to managing stress, emotions and overall health. As I mentioned in Part 1, I always used to suppress my emotions. But this technique has allowed me to be more accepting and playful with them. Beyond that, because this particular technique works on a cellular level,* the biological and physiological benefits are astounding. With disciplined and repeated practice, you can change the chemistry of your body. One of my clients recently had a shoulder injury, which meant they couldn't play for the season. While they remained on painkillers and continued their physio, I worked hard with them to accelerate the healing process. Every day we carried out creative visualisations to help their shoulder get stronger, alongside the medication. They would visualise knives stabbing their shoulder.

* Ephraim C. Trakhtenberg, 'The Effects of Guided Imagery on the Immune System: A Critical Review', *International Journal of Neuroscience*, Volume 118, Issue 6, 2008

That's what it felt like. One by one, we took the knives out, we visualised the cuts healing. They recovered far faster than expected and were back playing within two weeks!

Creative visualisation is incredibly personal, and there's no 'right' or 'wrong' way to do it. You want to embrace your imagination and let it take over. Whether your mind conjures up fantastical creatures or symbolic images, lean into it and let your body and mind communicate in their own unique way. The key is trusting the process and practising regularly to experience its full benefits.

CHAPTER 11

NEGATIVE VISUALISATION

If you prepare for the storm, you can dance in the rain.

Negative visualisation is the process of mentally rehearsing worst-case scenarios or things that might go wrong, such as financial struggles, losing a job or unexpected setbacks. By figuring out what to do in an action-oriented way, you become more prepared, resilient and, in some instances, ruthless when facing adversity. It has a long history, originating from the Cyrenaic philosophers before being adopted by the Stoics in 300 BCE. According to professor of philosophy William Irvine, it was the 'single most valuable technique in the Stoics' psychological toolkit' because it helped to reduce fear and enhance appreciation. Today, the application of the technique has largely changed and it is frequently used by elite athletes and CEOs, as well as continuing to be used by followers of Stoicism.

> **PROCEED WITH CAUTION**
>
> Negative visualisation is not appropriate for everyone's brain. According to Dr Bob Rotella and Dr Jim Afremow, the most common cause of choking in sport is imagining a negative outcome too often. This is even more likely if you already have a tendency towards negative thinking. The repetitive exercise will continue to wire your brain more negatively. If this sounds like you, for now, focus more on the other visualisations, but I would aim to still understand the process as it will help you in your day-to-day life to catch yourself if you are negatively visualising. On the other hand, if you think you might suit this type of visualisation, it can be extremely effective for being prepared, cultivating appreciation and even achieving your goals! When I work with clients, I spend time understanding their brain before suggesting they use negative visualisation.

ENHANCING PERFORMANCE AND BEING PREPARED

Michael Phelps famously used negative visualisation as part of his training and competitions. He imagined every possible scenario that could go wrong – his goggles filling up with water, losing his breath or falling behind everyone else. In fact, at the 200-metre butterfly final during the 2008 Olympics, Michael's goggles *did* fill up with water and he decided to swim 75 per cent of the race blind. To everyone's surprise, not only did he win, but he also got a new world record. In the interview after the race, he explained that because he had already played out the 'negative scenario' in his head, he handled the situation much better than he would have done otherwise.

In a more extreme example, Alex Honnold, the American free solo climber, was the first person to climb El Capitan without ropes or other safety equipment. This required rigorous mental training and preparation. As part of this, he would visualise every possible way he could die. Not only did this motivate him to keep going, but it also gave him the chance to prepare for the worst.

You can use the same technique to prepare for situations like going blank on stage, stopping during a marathon, losing points in a sports match, difficult conversations, redundancy or other challenges. Mentally rehearsing how you would react and how you would overcome it will undoubtedly put you in a better position.

Companies also use negative visualisation as a tool for future-proofing. Companies have specific teams who are looking at future scenarios that could pose a threat to the business. Scenarios like AI and robots replacing jobs or another pandemic are examples of this. They assess probability, action plans and crisis management. Teams can visualise these situations happening and figure out the best action plan. Doing it through mental rehearsal takes something theoretical and makes it more tangible. Leading psychologist Gary Klein designed a project management strategy embodying exactly this – called 'a PreMortem'. Before a project starts, the CEO or manager should ask the team to envision the project failing, seeing everything that's going to go wrong. He explains that too many projects or goals fail because of preventable reasons. It's not about thinking negatively, it's about being prepared for obstacles that will undoubtedly come your way.

PERFORMANCE AND PREPARATION VISUALISATION (2–4 MINUTES)

Recommended track: Either no music or 'Momentary' by Edvard Kravchuk

1. **Prepare:** Find a comfortable position and close your eyes. Move your body to release tension, take five deep breaths and a moment of stillness.
2. **Locate:** Start by mentally rehearsing the process or situation you are going into. Like always, add details. Where are you? What are you wearing? Build up the environment. What can you hear? Place yourself in the visualisation clearly.
3. **Visualise something going wrong:** Start visualising a likely issue or challenge. You can pick any scenario. Perhaps you start to lose the game, or you don't get the deal, or you see yourself getting anxious.
4. **Visualise overcoming it:** Now, start mentally rehearsing how you will overcome it. What is the best way you can respond? How will you handle it? See yourself bouncing back from the loss; show yourself how you will handle a rejection and overcome the anxiety in the mental rehearsal. You can try a few scenarios until one feels good.

5. **Add variety:** If you want to pick another obstacle, go for it! Each time, see yourself bouncing back, getting stronger and handling it in the best way you can.
6. **Close with resilience:** Take five more deep breaths, a moment of stillness and gently open your eyes.

ENCOURAGING APPRECIATION AND MOTIVATION

In this visualisation, you start by thinking about what you value in life – for example, family, health, house, friendships and so on. Then, you close your eyes and imagine life without them – if that thing or person was no longer in your life. Not only can this make you appreciate what you have in life, it may even change your attitude/actions towards it. For instance, imagining life without financial security might motivate you to save more or rethink your spending habits. Or picturing life without a loved one could encourage you to nurture your relationships and be more present. Fear of loss can be a powerful motivator for positive change.

You can also take this one step further by doing the deathbed visualisation. This is a technique that forces you to re-evaluate the decisions you are making. For example, let's say you want to quit your job and start your own business, but you are too scared to do it. Visualising yourself on your deathbed will give you a new perspective.

ACHIEVING GOALS

The Balcetis Lab carried out a study researching if people make more progress when they are visualising success or failure. The results showed that visualising the consequences of not reaching the goal nearly doubled the likelihood of achieving the goal. So far, the visualisations I have introduced you to have focused on positive outcomes, where we have imagined reaching our goal and connecting to the emotion of it. This study looked at goal-achieving and visualisation in a different way. By imagining what it would feel like to fail, and the possible negative consequences of falling short, the results showed that people became more motivated to push through discomfort and stay on track. Put simply, they were motivated by fear. This ties closely to the question that we explored in Chapter 7: 'Why do you want this goal?' What is the motivation behind it? Humans are generally better at moving away from failure rather than moving towards things we like or want, which is why this can be effective for some people. For example, if you were trying to quit smoking for your health and family, instead of visualising the reward for quitting, you would visualise the regret of not keeping the promise and suffering the consequences. It's not necessarily about being negative; it's about using a realistic but tough scenario to keep you disciplined with your goals. Try exploring both sides: visualise the success, but also spend a few moments visualising the consequences if you were to not follow through. How would you feel if that outcome actually happened? Does it motivate you more to keep going?

WHAT TO EXPECT

While I definitely see the value in using negative visualisation and there are brilliant studies to back it, I personally haven't found it to be as motivating in my own life. I asked a group of Olympians if visualising themselves failing helps them, and most of them agreed that visualising their success puts them in a better frame of mind. Again, this shows that visualisation is a tool to experiment with to see what works for you.

However, from working with my clients, I've found the most effective application of negative visualisation is using it to see yourself overcoming a challenge or obstacle. For example, I recently worked with the CEO of a tech company who wanted to improve his public speaking. We used process visualisation to see how he wanted to walk, talk and deliver, but added in a negative visualisation where he went blank in the middle. This feels so real and can be uncomfortable for people, but he saw himself overcoming it by taking a pause, adding a little joke and then carrying on. Two months later, he was on stage and a similar incident happened. But instead of panicking, he did exactly what he had mentally rehearsed. Even though he went blank on stage, it didn't affect his performance at all. Using negative visualisation as a way to build extra resilience and prepare for scenarios is extremely effective.

CHAPTER 12

EXPLORATIVE VISUALISATION

Logic will get you from A to B.
Imagination will take you everywhere.
– Albert Einstein

Explorative visualisation is probably the most undocumented and unknown technique. It can also be referred to as receptive visualisation. Unlike goal-driven visualisation, this technique is about embracing the vast, boundless realm of imagination. Where creative visualisation uses the imagination specifically for emotions/the body, explorative visualisation uses it to explore ideas, solve problems or create entirely new concepts – think of it as a mental playground. With this approach, you're not confined to any specific outcome. Instead, you allow your imagination to unfold freely, unlocking levels of creativity you may not even know you possess. In some instances, you also want to get curious about how your body reacts to the images created.

I started experimenting with this technique at university. I studied Geography with Innovation and a big part of my degree required coming up with new business ideas, brainstorming and also pitching. I remember there was a particular presentation coming up for an exam and I didn't really know where to start. I closed my eyes and saw different ways I could do it. Some ideas were pretty crazy and random. Some I liked and resonated with. By the end of the visualisation, I had the exact structure of the presentation, the themes I would talk about and what my slides would look like. I then took what was in my mind and made it happen. Naturally, a lot of us do this anyway as we have a general idea of what we want to create in our mind. But when you make the process even more intentional, the level of creativity and ideation is better. So think of explorative visualisation as thinking on steroids!

After practising this technique enough times, I did my own research to see who else was doing it and I wasn't surprised to find that it was used by some of history's greatest minds. The first time I saw someone articulate this idea was in Benyamin Cohen's book *The Einstein Effect*. Einstein called this 'mental imaging' and used it to access his brain's natural imaging abilities, cognitive flexibility and enhanced creativity. I also discovered, through reading his biography, that Walt Disney practised the technique throughout his life. It has been widely documented that Walt Disney's visualisation ability was sensational. He called it 'imagineering'. According to his personal diaries and work colleagues, Disney had the remarkable skill to visualise multiple perspectives, solve complex problems and design unique characters, all in his mind. This process involved him intentionally taking time to close his eyes and mentally create.

As well as Disney, some of the greatest thinkers, inventors and creators have also used this – including Einstein and Nikola Tesla. Tesla is best known for inventing the electric coil, the mechanism behind the Tesla cars we see today. But the thing no one really mentions is that he used visualisation to do exactly this. He stated that he 'needed no models, drawings or experiments. [He] could picture them all as real in [his] mind.' In other words, Tesla was using explorative visualisation to take advantage of a highly developed version of a mental ability that we all share.

CREATING CONTENT

While I am no Einstein or Walt Disney, I consistently write seven LinkedIn posts and create five Instagram reels a week using explorative visualisation. For LinkedIn, I will think about the topic I want to write about, then I will close my eyes. Let's take negative thinking, for example. I picture a white blank screen and use my mind to write out some variations of the post. I then start editing them in my mind. I go as detailed as seeing myself deleting words or reading it out mentally. Once I open my eyes, I then write them out on my computer. Of course, there will still be some level of editing needed, but an hour of work becomes ten minutes.

With Instagram, I pick the music I am going to use for a specific video. I then close my eyes (while playing the music) and storyboard the content. I try out different words, ideas and scenes. Once I have created something in my mind, I find it very easy to create on my phone. Like I mentioned before, you probably do some version of this already, but when you intentionally close your eyes and visualise, the impact is much greater as you are tapping into a part of your brain that is more creative and efficient. One of my clients recently told me they were doing a big presentation for their company, but they were having a huge creative block and finding designing slides draining. Instead of physically doing it, I guided them to close their eyes, see the blank PowerPoint slides and start creating them in their mind: see the diagrams they wanted; see the text on the pages; see the flow of the presentation. When we had finished, they immediately jotted down all their ideas and had nearly everything they needed to get it done.

Even this book is largely a result of visualisation. I have never written a book before and struggled with how I wanted to structure it. Following the same process, I created a flowchart in my mind and kept swapping out different sections. Afterwards, I wrote them all on Post-it notes and put them around my room. In fact, as I have been writing it each day, there are some chapters I get a bit more stuck on. The thing that is helping me the most is closing my eyes and just writing it in a visualisation. When I open my eyes, I have so much more I can write, simply because I took the time to use my mind slightly differently.

So when you are creating any type of content, I would suggest starting with a loose theme or idea. Tesla mastered this capacity in his brain and it made him the genius he was. As he wrote in his autobiography, *My Inventions*, 'When I get an idea I start at once building it up in my imagination. I change the construction, make improvements and operate the device in my mind. It is absolutely immaterial to me whether I run my turbine in thought or test it in my shop.'

It's time to embrace your inner Tesla! Let's start really simple. Imagine you want to write a tweet about how to become confident:

CREATING CONTENT VISUALISATION (2–3 MINUTES)

Recommended track: No music needed

1. **Prepare:** Find a comfortable position and close your eyes. Move your body to release tension, take five deep breaths and a moment of stillness.
2. **Locate yourself:** Start by imagining you are in front of your computer screen.
3. **Mentally type:** Begin typing a tweet about becoming confident. Start anywhere. A word, a quote, an idea.
4. **Mentally edit:** Keep typing, deleting and editing.
5. **Mentally read:** Finish the tweet and read it in your mind a few times.
6. **Close with creativity:** Take five more deep breaths, a moment of stillness and gently open your eyes.
7. **Create:** Now, write down what you visualised.

Some of you may have found this harder than others, and that's normal. This technique took me the longest to master. But once you do, your imagination becomes a genius tool in work and life.

DESIGNING FOR SPEECHES, EVENTS AND PRESENTATIONS

Since university, I have continued to apply this technique to my events and keynotes. For example, when I got asked to do a TEDx Talk, I was told the theme and that it would be 20 minutes. Before putting pen to paper, I scripted the entire talk in my mind first. I carried out the exercise a couple of times because it was a much longer talk than normal. But exploring the different options gave me all the information I needed.

Another example is Mental Fitness Live. A year ago, it didn't exist. My co-founder and I knew we wanted to run an event, but we had never done it before so we were starting from a blank slate. We didn't know how it should look. This was the perfect time to use something like explorative visualisation. I closed my eyes and began designing the event in my mind. I imagined everything: the staging, the audience, the content, and even the small details. As I explored different possibilities, I noticed ideas that I hadn't initially considered. Eventually, I had a clear vision of what I wanted, which helped us secure a location and launch the event within a week. As I continue to improve and refine these events, I rely on this visualisation for both big and small ideas. Last month it helped me to think of a specific song to play as people enter, and another time it gave me the idea to invite a special guest on stage. The skill of visualisation allows me to create and deliver truly unique experiences that stand out in a crowded field.

Our conscious minds can limit us, so when we tap into the creativity of our brain, new possibilities arise. I want you to think of an upcoming event or speaking opportunity where you could benefit from getting a little more creative or organised with the planning. Following a similar process, close your eyes and let your imagination do the hard work. Explore the vision and details, and allow this to be your guiding compass.

MAKING DECISIONS AND SOLVING PROBLEMS

A couple of years ago, I met Gaby Mendes, host of the *Talk Twenties* podcast. When we first chatted, she was in the middle of making a big decision for her company. It was a decision that would change the course of her life, but she didn't have a clear answer yet. I told her I would record a visualisation that would help her make the decision. When it comes to making decisions, it's impossible to predict exactly what will happen, but there are definitely things we can do to help, especially with big ones. In the visualisation, I asked her to imagine that she had picked option A. Then I guided her to explore what the company and her life would be like having taken that option. Where would she be working? Who would she be with? What would her daily activities be like? I asked her to stay very aware of how her body felt and any reactions that came up. This is important because the body will tense or close up if we don't feel aligned to something. Then I asked her to pick option B and guided her through the same process.

This allowed her to explore both options in a different way. Instead of writing a pros and cons list, she immersed herself in the reality of it. A few weeks later, she messaged me saying that the visualisation made her realise which decision she wanted to go for. Since then, she has told me she still thinks back to this moment as it really did change the course of her career and life.

Similarly, when we are faced with problems, we often get overwhelmed and consumed. The more we talk about it, the more stressful it becomes. It can help to take a step back and approach it in a different way. I learned an extension of this technique in the book *The Einstein Factor* by Richard Poe and Win Wenger. They talk about the power of thinking in metaphors. Using metaphorical thinking can help us approach problems from different angles, leading to better solutions. Metaphors bridge the gap between conscious and subconscious thought, making abstract concepts more understandable. So, for example, if you are feeling stuck in life, you may visualise a train getting stuck on the tracks. In the visualisation you may try to push the train, fix the engine, get some help or pick a different track. Maybe you get rid of all the people on the train and then try to push it. This can give you a new sense of perspective when approaching your actual problem. Maybe it makes you realise that you need to reach out for help. Or that you need to let go of certain people weighing you down. The options become endless. In some ways, this combines creative visualisation with explorative visualisation as we are using imagery to make a situation more tangible, but are using exploration to find some new options. Next time you are making a decision, try mentally rehearsing each option. Immerse yourself in what your life would be like if you were to choose one way or the other. You want to become aware of how your body feels and the thoughts occurring in your mind. This will be the signal or data to listen to.

GENERATING NEW IDEAS AND CREATIVITY

One of my favourite examples of explorative visualisation comes from Bill Russell, 11-time NBA basketball champion. Russell was one of the best defenders the game has ever seen, firstly because of his performance, but secondly because of his creativity on the pitch. Russell was always looking to be better and change his game. He would spend time closing his eyes creating new defensive moves he could try in practice. He said, 'I was happy because the defensive moves were the first that I'd invented on my own and then made real. I didn't copy them; I invented them. They grew out of my imagination.' Your imagination is like a science lab; you can run experiments without the constraints or negative repercussions.

Now more than ever, companies are under pressure to think of the new best thing or create something out of this world. But if we keep thinking in the same ways, we will keep getting the same results. We are good at sticking to the same way of doing things – the same brainstorming sessions, the same results. But the mind is an incredible tool for inventing new concepts, coming up with ideas and getting creative. Recently, I worked with a company

EXPLORATIVE VISUALISATION

that was changing its retail landscape. It was starting from scratch. I was actually called there to deliver a mental fitness workshop, but I stayed for the rest of the sessions. I was sitting in the ideation phase for the new retail stores and people were presenting their ideas. Honestly, they were all quite boring.

This company's brand was built on cutting-edge technology and innovation, yet they were coming up with the same concepts as everyone else. I took a bold move and suggested to the CTO that I try one exercise with everyone. He agreed. I asked everyone to sit down and close their eyes. I started the visualisation by asking them to stand in an empty shop – white walls, lots of space and nothing inside. Then I asked them to go as bold and brave as they could, designing the retail stores, seeing what the products would look like, creating unique events, seeing how people moved through the shop, the sounds, the music, all of it. We did this for about seven minutes and then I asked them to open their eyes. Using what they saw or created, they jotted down some ideas and made them more tangible. Twenty minutes later, they presented new concepts and the ideas were remarkably different. Yes, some of them were extremely 'out there', but at least the team was thinking bigger and better. They were going beyond their normal way of operating and that was exciting for everyone. The CTO came up to me afterwards and told me he had never seen his team work in that way. I told him it wasn't magic – it was mental fitness.

You can try a similar process for your own life or work. I have outlined a visualisation for creating new ideas on the following page.

GENERATING NEW IDEAS VISUALISATION (5 MINUTES)

Recommended track: Either no music or 'Epic Emotional' by AShamaluevMusic

1. **Prepare:** Find a comfortable position and close your eyes. Move your body to release tension, take five deep breaths and a moment of stillness.
2. **Visualise an empty canvas:** Picture yourself standing before an empty canvas, an open space or a blank page. This represents the potential of the creative process – everything is possible, and there are no limits. The emptiness of the space invites new and different ideas to flow.
3. **Brainstorm ideas:** With the particular focus/theme you are using, allow ideas to begin flooding your mind. No matter how wild or unconventional they may seem, let them come in without editing or judging. Picture the ideas forming as images, scenes, words or symbols.
4. **Door analogy:** You can take this further by imagining different doors. Each door you enter is a new idea. Go deeper into each possibility, experience it and notice how you feel. Tune into how your body reacts and where your mind goes.
5. **Close with clarity:** Take five more deep breaths, a moment of stillness and gently open your eyes. Write down or chat about what you saw. Decide if you want to take action on an idea or simply reflect on some of the directions.

WHAT TO EXPECT

The immediate effect of explorative visualisation is extremely varied. For some people, it feels like they have just downloaded lots of ideas. For others, they feel excited and have gained clarity over a problem. Some people also feel more confused. I've noticed that creatives, visual thinkers and inventors resonate with this technique a lot and, when they integrate it, they see a considerable difference in how they approach their work and life.

The more you do this one, the better you get at it. Because there is very little guidance, the mind likes to do its own thing, so it requires you to really focus, but also know what to focus on. Lots of different images and ideas come up, so knowing which ones sit well with you or don't is important. I'd recommend testing this out on small things first. In the long term, explorative visualisation will help you tap into a part of your brain that 99.99 per cent of people never will (unless they read this book).

I know it might seem like there's a lot to practise now that you're familiar with so many visualisation exercises, but my goal was to provide you with all the tools and insights in one place. This way, you can easily refer back to any exercise whenever you need it. Remember you can return to any section of the book whenever you need practical guidance or are facing a particular hurdle in your life.

In Part 3 I will show you how to incorporate these visualisation techniques into your everyday life and build a routine that works for you. Let's dive in.

PART 3

EXECUTION: PURSUING YOUR EXCELLENCE

Excellence is not being the best; it is doing your best.
— *Robin Sharma*

So far you have learned the theory behind visualisation and have had the chance to explore the different techniques in practical and applied ways. You might be thinking, where do I begin with this? This section will give you the tools to not only start your visualisation journey, but also turn it into a daily practice that transforms how you think, feel and perform. Doing a single visualisation can easily give you short-term benefits. But true mental fitness comes from sustainable routines and discipline. As Billie Jean King said, 'Champions keep playing until they get it right.'

CHAPTER 13

BUILDING YOUR ROUTINE

> *I am not a product of my circumstances. I am a product of my decisions.*
> – Stephen R. Covey

Building a consistent and regular visualisation routine is key to reaping its full benefits. To make it most effective, it's essential to establish a routine that works for you – one that can integrate seamlessly into your day and become a non-negotiable part of your mental training. The point isn't to create a perfect routine, the point is to create a sustainable one. In this chapter, we'll explore the best times, ideal environments and recommended durations to practise visualisation so you can create a personalised routine that maximises its potential.

WHEN

There's no hard rule for when to practise visualisation, but certain times can be more effective due to how our brain functions.

If we look at the five different techniques we explored in Part 2, here are the general suggestions I would give:

- Outcome visualisation: morning or just before bed
- Process visualisation: morning, just before tasks/performance or just before bed
- Creative visualisation: anytime
- Negative visualisation: anytime
- Explorative visualisation: anytime

In the morning, your brainwaves move from theta (deep relaxation) to alpha (relaxed alertness). This is a time when your mind is most suggestible, meaning it can absorb new ideas, emotions and goals without getting distracted by the usual noise of daily life. If you wake up, immediately look at your phone, watch the news or get overwhelmed with emails, your brain is absorbing and rehearsing negative ideas, stress and worries. Instead, when you feed it your goals, your vision, or mentally rehearse who you want to be today, you shift the quality of your thoughts and emotions considerably. Take a moment to ask yourself: How did you start your day today? Now, imagine how different you would think and feel by starting it with a primed and purposeful mindset. Further, most people prefer to practise visualisation in the morning because it's like a blank slate. You are rested, getting ready for the day and can choose how you want to show up. And of course, if you are using process visualisation to rehearse your day, then morning is the best time. Not only will this give you an immediate boost or sense of relaxation, but it will also prepare you for the day extremely well. Personally, on days when I need extra motivation or focus, I revisit my visualisation practice throughout the day.

Some people prefer doing their visualisations just before bed, and there's a good reason for this. When you focus on your goals or performance right before sleep, your brain continues to rewire itself while you rest. Neuroplasticity – the brain's ability to reorganise itself – primarily happens while you sleep, making this a win-win scenario. Plus, when you're tired, your brain naturally shifts into alpha waves, which makes it easier to access your subconscious mind. Even if you fall asleep during the process, a few minutes of active and intentional visualisation can still help train the mind and yield benefits. A guided recording can be especially helpful if you find it difficult to stay awake during your visualisation.

Personally, I've found that high-energy goal visualisations before bed don't work for me. I end up feeling too energised and eager to take action. But performance and character-based visualisations help me wind down and sleep better.

As you build your visualisation routine, especially for outcome and process-based visualisations, consistency is crucial. It's like brushing your teeth: if you did it at random times, you'd probably forget, right? Try to pick the same time each day, especially in the beginning. Once the habit is ingrained, you can be more flexible with timing.

Creative visualisation, on the other hand, can be used whenever you need it. For example, if you suddenly feel angry or jealous, you can quickly do a few minutes of visualisation to reset your mental state. However, think of creative visualisation not just as a reactive tool, but as a preventative one. When practised regularly, it helps you manage emotions before they escalate, fostering a sense of calm and control.

In the same vein, negative and explorative visualisations can be beneficial, but how often you practise them depends on your individual needs. Some of my clients use negative visualisation a week before a performance to remain agile, but stick to positive process visualisation nearer to the actual event. I've also worked with people who use the deathbed visualisation (see page 175) weekly, finding that it helps them appreciate the present moment and gain perspective on their lives. Likewise, clients in the creative industries or those working on large projects often use explorative visualisation daily to refine their ideas, enhance creativity and solve problems more effectively.

When establishing your routine, you want to choose one or two visualisations to repeat regularly. This ensures that you build a strong foundation for mental fitness. Then, depending on what's happening in your life, pepper in the other visualisations – whether that's for managing stress, enhancing creativity, or refining your goals.

WHERE

Ideally, you'll want to practise visualisation in a quiet, undisturbed space. It's the easiest way to avoid distractions, especially when you're just starting out. But the beauty of visualisation is its versatility: you can do it anywhere. I've visualised on the Tube, in parks, and even perched on a busy bench in Oxford Circus. Sure, people stared – but why should mental training be any less normal than jogging in public?

HOW LONG FOR

When I first started practising visualisation, I managed just two minutes, then five, and eventually worked my way up to ten. Now, I practise for an hour a day – sometimes even two. This isn't something I feel I *have* to do; it's something I genuinely *want* to do. Of course, there are days when I only manage a few minutes, and that's okay. But after nearly a decade of consistent practice, I've built up my mental endurance, much like an athlete gradually increasing the weight they lift. Over time, my mind has grown stronger – it can mentally rehearse, create and process more than ever.

Currently, my routine includes eight different visualisations, stacked on top of each other. I start with two creative visualisations (one for relaxation and one for health), followed by three outcome-based visualisations (big vision, character and short-term goals). Next, I do two process visualisations (performance and daily routines), and I finish with one final creative exercise (gratitude). But this isn't something I suggest for beginners. When you're starting out, stick to one or two visualisations and practice for seven to twelve minutes, at least three times a week. In Part 2, I've suggested ideal durations for each type of visualisation – use these as a guide, but feel free to adjust based on what works for you.

I guide a well-known visualisation at my Mental Fitness Live events, and once it's complete, I ask participants how long they think the exercise lasted. Most people guess around five minutes, with some estimating just 90 seconds. Then, I tell them: it was 14 minutes and 48 seconds.

They're always shocked, because they never imagined they could sit and visualise for that long. But time and time again, I've seen just how possible it is – no matter where you're starting from.

Above all, the key is repetition. Practising for just two minutes a day will yield more benefits than an hour-long session once a week. Think of it like physical fitness: going for a run once feels good, but it's the consistency that builds endurance and improves your fitness. The same principle applies here. Consistent, repeated practice is what leads to lasting change. If you already have an established routine or are enjoying your practice, you can gradually extend the length and frequency of your sessions.

TAILORING YOUR VISUALISATIONS

Have you ever heard of Build-A-Bear? It's a children's store with a fun concept where you take a plain teddy bear, personalise it, dress it up and make it entirely your own. I want you to treat your visualisation and mental fitness in the same way.

Think of it this way: when you go to the gym, do you follow the exact same routine as everyone else? Of course not. You have a unique body, distinct goals and specific needs. It takes time and experimentation to discover what works for you, and the same is true for mental training. The key is to craft a routine that feels like yours.

For example, you might start with two minutes of breathwork to centre yourself, followed by four minutes of meditation, seven minutes of visualisation, and a brief stretch or light movement to close. Or perhaps you do ten minutes of visualisation right after a morning workout, where the endorphins enhance your focus and creativity. Alternatively, you might flip the order and start with mental training to sharpen your mindset and motivation before exercising.

The point is to tailor your practice to your needs. One of my clients, an entrepreneur and avid tennis player, found that while visualisation improved his game, he still felt jittery before matches. When he shared this with me, I suggested extending his breathwork after visualisation to calm his nervous system. That simple tweak transformed his pre-match confidence. If you're just starting out, guided recordings can be incredibly helpful. They provide prompts to keep you focused and help direct your mental energy effectively. Even now, I still use my own recordings to stay on track. As you gain experience, you might transition to self-guided sessions or experiment with customising your practice.

Remember, while the core principles of visualisation stay the same, your practice is yours to shape. Personalisation isn't just encouraged – it's essential. Pay attention to what works for you, adapt as needed, and let your routine evolve alongside your goals.

BUILDING A VISUALISATION HABIT

Building a habit of visualisation can feel overwhelming at first – there's so much advice out there, often conflicting, about what to do, when and how. But the process doesn't need to be complicated. Let's break it down simply, with the same principles I've followed to develop a consistent practice over the years.

At its core, a habit is an action repeated regularly until it becomes second nature. However, this doesn't happen overnight. As Robin Sharma wisely puts it, 'All change is hard at first, messy in the middle, and gorgeous at the end.' So, how do you push through the messy middle and make visualisation a habit?

The first step is to connect with your **why**. Why do you want to visualise? Is it to achieve your dreams, improve your mental fitness, or become the best version of yourself? The 'why' needs to be meaningful, not just 'it's good for me' – that's rarely enough motivation to stick with it. Your 'why' is what will pull you back to your practice during the inevitable moments when you feel like quitting. It grounds you and reminds you of what you're working toward. And remember, you don't need to be unhappy to want to improve.

Once you have a strong 'why', it's time for the next step: **take action**, but don't go overboard. One of the biggest mistakes people make when starting a new habit is jumping from 0 to 100. I don't expect you to start visualising for 30 minutes a day right after reading this book. As I said above, it's better to start with a few minutes and then build from there. Think of it like each action you take is a little vote for yourself and your goals. The aim is to start increasing the number of votes you give yourself.

BJ Fogg's Tiny Habits® method is a great guide here. The first step is to start small. Actually, tiny. You want to make the behaviour or action so small that it is almost effortless. For example, if you wanted to build the habit of reading every day, you would start by reading a paragraph of this book every day. Yes, that small! By reducing the behaviour to a tiny, non-intimidating task, it becomes very hard to not succeed – even when motivation is low. The next step is to anchor the activity to an existing habit. This makes it easier to remember and integrate into daily life. For example, before brushing your teeth, you read a paragraph of the book. The last step is to celebrate. Remember the little victories that we talked about in Chapter 6? After you read the paragraph, immediately reward yourself and celebrate. This could be as small as saying 'Yay, nice one' or even just a smile. Celebrating after completing the small habit might seem unnecessary, but it creates positive reinforcement, so the likelihood of you repeating the visualisation increases. You'd be surprised how quickly a new habit can form when you have a positive emotion associated with it. Fogg outlines a simple Tiny Habits® Recipe:

> After I [existing habit], I will [new tiny habit].
> Then, I will [celebrate].

Here is an example:

> After I brush my teeth [existing habit], I will do
> two minutes of visualisation [new tiny habit], then
> I will say 'go you!' [celebration]

Over time, the tiny habit will become more automatic. Naturally, the length or size of the habit will increase. For example, someone who starts by doing two minutes of visualisation might eventually start doing ten without much extra effort, because the initial habit is firmly established.

YOU REAP WHAT YOU REPEAT

In Part 1, I explained how you are currently reaping everything you are repeating: thoughts, beliefs, patterns and emotions (see page 51). They have been repeating for years and have shaped you and your worldview. As you've probably guessed, the solution is in the problem. Because you reap what you repeat, you've got to change what you repeat. That's why consistent visualisation is the key to mastering the skill and reaping the benefits. Mastery requires practice. There is no shortcut.

I want to use physical fitness to emphasise my point clearly. It is impossible to get abs after doing 100 sit-ups. Yes, you might feel good and strong after, but for long-term change, you need to repeat sit-ups consistently throughout the week. The same goes for mental fitness. When you first start thinking new thoughts or visualising new behaviours and performances, it might feel pointless. The brain might even resist your new practices. And you might not believe them, especially if there is a big gap between your visualisation and reality. When you keep doing them, though, the thought and belief get deeper and deeper. The brain becomes more familiar with it and that becomes your new city (see page 30).

Here is the thing no one tells you: visualisation by its nature can be very energising and powerful. Especially at the beginning, you may feel on top of the world or you might experience intense emotions. But sometimes I get messages from people saying, 'I keep doing this visualisation, but it's not having the same effect on me anymore.' Just because it doesn't have the same novel intensity doesn't mean it's not working. The more you practise something, the more routine it becomes. But if we are really clever, we will find ways to stay enthusiastic and interested in the fundamentals. I remember watching an interview with Kobe Bryant where he explained that the thing that gave him the edge over other players was the fact that he never got bored of the basics. In fact, he committed to becoming world-class at them.

THE PAINKILLER VERSUS SUPPLEMENT MENTALITY

I want to introduce the painkiller versus supplement mentality to help you think about how you approach mental fitness. A painkiller addresses an immediate need. If you visualise when you're stressed or overwhelmed, it's like taking a painkiller – you're treating the symptom in the short term. Yes, it can give you relief, but it won't address the root cause.

A supplement, on the other hand, is a proactive, ongoing practice that enhances your overall health and performance over time. It's about committing to your mental fitness as a daily habit – whether you feel great or not. This long-term approach creates lasting change and in some cases prevention, rather than just providing temporary relief. When it comes to visualisation, both approaches provide value.

When I started visualising, I used it as a painkiller – a way to 'fix' myself when I was feeling ill and hopeless. But I didn't stop once I started feeling better. As I kept practising, visualisation helped me tap into a level of performance and confidence I had never experienced before. It became my daily mental 'supplement'. Sometimes the exercises will help you cope. Other times they will help you thrive. Sometimes they will calm you down. Other times they will ignite a powerful fire within you. Sometimes, they will elevate your performance to new heights. Other times, they will reveal a strength you didn't even know you had. Investing in your mind when you are feeling good is just as important as investing in it when you aren't. That's where you can really meet your potential.

This goes beyond just individuals. Often when I first start working with companies, it's clear they have adopted the 'painkiller' mentality. They wait till their workforce is burned out before they take action. And when things are going well, employee well-being is the first to be sacrificed. Yet human capital is a company's most valuable resource.

So why is it not prioritised? Why don't we give teams the best chance to invest in their mental fitness and pursue excellence? High-pressure environments demand peak performance, focus and resilience. These qualities can't be achieved with a reactive, painkiller mindset. They require consistent, steady supplementation – a daily investment in mental fitness that prepares employees to excel, not just endure. The lesson is clear: whether you're an individual or a company, the key to thriving isn't reacting to problems. It's building a strong foundation that allows you to rise to every challenge.

ADD SOME TOPPINGS

Over the years, I have started to add other things to my visualisations, mainly journalling, vision boards and affirmations.

Some of you might be rolling your eyes seeing these words, but hear me out.

Journalling

Journalling is a great practice to do both before and after your visualisations. Writing down your thoughts and goals can bring immediate clarity and focus, especially when you're juggling multiple ambitions and feeling uncertain about where to begin. According to Dr Gale Matthews, a psychology professor at Dominican University in California, people who wrote down their goals were 42 per cent more likely to achieve them compared to those who didn't write their goals down. A powerful exercise is to first list both your personal and professional goals, then prioritise the ones that matter most to you right now or need the most attention. Journalling can also help clear mental clutter. If you're feeling overwhelmed or frustrated, jumping straight into a visualisation might amplify those emotions. Instead, take a moment to write down your thoughts. This can help you feel more grounded, making your visualisation more effective.

Conversely, doing some reflective journalling after your visualisation lets you track your progress, note insights and refine your approach.

Vision boards

I used to be skeptical about vision boards. How could a few pictures really help me achieve my goals or improve my mindset? And let's face it, there's often a stigma attached to them. People think vision boards are all about flashy cars, piles of cash or dream homes. But the neuroscience behind them is far more compelling than that.

A vision board is a collection of images or drawings that represent your future goals and aspirations. Whether physical or digital, these visuals work because the brain assigns higher value to the images it sees repeatedly. This is called 'value tagging', a concept I learned from neuroscientist (and author of *The Source*) Tara Swart. The more you engage with the images, the more your brain views them as important. This primes your brain to notice opportunities aligned with your goals, thanks to the Reticular Activating System (RAS) (see page 28). One of the pictures that has been on my vision board for over five years is a healthy pink colon that I drew myself.

That said, a vision board alone isn't enough. To maximise its power, I combine it with visualisation and daily action. It's one thing to look at a goal, but it's another to mentally rehearse it and take consistent steps towards it. True transformation happens when you wire your goals into your brain and live them daily.

Affirmations

Affirmations often get a bad reputation, but the world's top performers swear by them. Whether they call them 'power phrases', 'trigger words' or just 'positive self-talk', affirmations have been proven to be beneficial for the mind and body.

Take Cristiano Ronaldo, for instance – before stepping up to take a penalty, he repeats, 'You can, you can, it's normal for you to score.' Or Muhammad Ali, famously declaring, 'I am the greatest' before he actually was. These phrases get athletes into the optimal mental and physical state to think and perform at their best.

When it comes to your self-talk, there are three types of affirmations to incorporate into your routine:

- **Motivational affirmations**: Statements like 'I can do this' or 'I'm giving it my all' boost your confidence and energy.
- **Instructional affirmations**: Simple phrases like 'Walk, pause, smile' provide clear cues that enhance focus and competence. (This is where the T-CUP formula on page 135 comes in.)
- **Character-based affirmations**: These begin with 'I am' and reinforce your identity. For example, 'I am confident' or 'I am ready.'

I used to repeat 'I am healthy' in my visualisations when I was unwell. At first, I didn't believe it. But the more I said it, the more my mind started to internalise it. You might find it awkward at first, but think about this: what's the harm in trying? With everything you've learned about how beliefs form in the brain, it makes sense to input helpful ideas. Yes, it might feel a bit 'fluffy' or even cringey at first, but when you start to feel it, repeat it and eventually believe it, affirmations can serve as a powerful catalyst for rewiring your brain.

When selecting your power phrases, make sure they feel **personal, positive,** and **present**. Depending on the situation, your affirmations might change. For example, one study found that the top five tennis players in the world set themselves apart with their use of affirmations *after* making mistakes on the court. Instead of focusing on the error, they reinforced what they were doing well, which helped calm their heart rate and breathing, allowing them to quickly recover and continue performing at a high level.

Being aware of the self-talk you need when you're feeling good, neutral, or low is crucial. Your relationship with your self-talk evolves over time, and learning to manage it will set you up for success.

There are many ways to integrate affirmations into your practice. Some people write them down, others speak them aloud; I personally repeat them in my mind during my visualisation sessions. This has two benefits: first, it links specific words to the imagery in my mind, strengthening the connection. Second, it supercharges my mental energy and focus. I don't use affirmations constantly, but when I do, they provide the exact boost I need.

Incorporating these 'toppings' into your visualisation routine brings depth and versatility to your mental fitness journey. You might be doing these already, which is great. Allow them to amplify your efforts and accelerate the rewiring process.

CHAPTER 14

FALL IN LOVE WITH DISCIPLINE

Make choices today that build the life you envision tomorrow.

Over the years, I've devoted a lot of time to mastering self-discipline. Partly because I had to, due to the illness, but later, because I realised it's the highest form of self-love we can give ourselves. The truth is, discipline itself isn't the hard part – *consistent* discipline is. That's because discipline is a perishable skill. The moment we fall off track, it's all too easy to stay off. It takes character and courage to get back on again.

True discipline isn't the one we experienced at school, where someone tells you what to do and then you get punished if you don't do it. Real discipline is when you align your intention, action and execution – from thinking, to doing, to being. With time, I have built discipline in many areas of my life – my food, my boundaries, my visualisation practice and my goals. It may sound counterintuitive, but discipline isn't about restricting yourself – it's about giving yourself the freedom to do what truly matters. Through discipline, I've created the space, energy and time to live the life I've always wanted.

I follow seven key rules when it comes to maintaining discipline in both my life and my work. The rules I'm about to share won't just help you cultivate discipline in your mental training – they'll improve every area of your life.

Here are the rules:

RULE 1: MAKE IT PERSONAL

The first question to ask yourself is why? Why are you trying to become disciplined? Why are you trying to maintain a habit? Who are you doing it for? I made visualisation personal to me. It became part of my own mission to become healthy and take control of my life. I remember saying to myself, what if you decided to really give this a go? Make it happen for yourself. And, well, the rest is history. Making it personal pushes you to go that one step further for yourself. Do you want to feel better? Do you want to level up? Do you want to progress in your career? Do you want to manage your emotions better? Do you want to be a better colleague, sibling or partner? **Discipline is personal**.

RULE 2: WORK AGAINST THE HACK

Hacks have killed human discipline – 'Lose weight in thirty days', 'Become a millionaire in three months', 'Three hacks to become super confident'. I even saw one that said 'Cure ulcerative colitis in two weeks.' Of course, I signed up – that was the dream! We are lured into 'hacking' everything these days, especially when it comes to our health. We want the quick fix. And the sad thing is, so many people are buying into them. As tempting and enticing as they are, there is no elevator to success. You can't just skip steps, especially if you want long-lasting results. You have to take the stairs and learn to enjoy or embrace each one, no matter how big or small. So instead of trying to hack your life, hack your brain! Visualisation is not a quick fix when it comes to training the mind and it won't transform your life overnight. But making it a part of your life, just like physical fitness or walking, will undoubtedly bring you incredible results. **Play the long game. The rewards will follow.**

RULE 3: SHOW UP FOR YOURSELF EVERY DAY

One of the quickest ways to lose discipline is to put people or things before yourself. It's a slippery slope: you skip a workout to finish a colleague's task, skip meals to care for your family, or push your own needs aside to meet a deadline. Before you know it, your physical, mental and emotional health starts to unravel.

The solution is simple but profound: make it non-negotiable to show up for yourself every day. This doesn't mean grand gestures. It can be as simple as taking one minute to breathe, stretch or set an intention.

Take Sarah. A single mum, corporate executive and self-confessed perfectionist, she came to me exhausted. Her mornings were consumed with her kids, her days revolved around work and her nights ended in sheer exhaustion. She said, 'I don't even know what showing up for myself looks like anymore.'

We began with five quiet minutes each morning, no rules, no pressure. Some days, she meditated. Other days, she sipped coffee in silence or jotted down a quick note in her journal. The first week felt awkward – she wondered if it was enough. But over time, those five minutes became sacred.

Gradually, Sarah noticed shifts: she had more patience with her kids, more focus at work, and even began adding longer walks to her routine. What began as a small, simple act of self-care rippled into every aspect of her life. She told me, 'It feels like I'm filling my own cup for the first time in years.'

Not every day will feel like a victory. There will be days when showing up feels impossible – when you're tired, overwhelmed or unmotivated. On those days, just show up. Even if it's just for a moment.

The act of showing up, no matter how small, builds momentum. The results may not be immediate, but they are inevitable. So, start today. Practise prioritising yourself, even if it's for just one minute. **It's the small moments that lead to lasting transformation.**

RULE 4: LEARN TO LOVE AND COMMIT TO THE PROCESS

A few years ago, I worked with a client named James, an entrepreneur who was training for his first marathon. When we began, he was full of energy, eagerly mapping out routes and buying all the latest running gear. But as the weeks wore on, the excitement faded. The early-morning runs became gruelling. Progress felt slow. One day, James vented to me, 'Why am I even doing this? I don't feel like I'm getting anywhere.'

I reminded him of something simple: progress is often invisible in the moment. The real transformation isn't just in crossing the finish line – it's in every step leading up to it. James didn't need a breakthrough; he needed a mindset shift. He started to focus on showing up for the run itself, not just the end goal. Slowly, he found joy in the process.

The truth is, the process can be messy, boring and exhausting. But it's also where the magic happens. There's a Buddhist teaching that captures this beautifully: 'Before enlightenment, chop wood, carry water. After enlightenment, chop wood, carry water.' It's a reminder that whether you're just starting or achieving mastery, the work remains. The process is never finished – and that's its beauty.

This rule asks you to commit. To give your time, energy and sometimes money – even when the results aren't immediate. With visualisation, I often see people make a critical mistake: when things get good, they stop. They assume they've 'arrived' and no longer need to invest in the process. But mental fitness is not a destination; it's a lifelong practice.

James eventually finished his marathon, but what stuck with him wasn't just the medal. It was the discipline he built along the way. He told me later, 'I thought I was training for a race, but really, I was training myself.'

Whether you're feeling good or feeling bad, continue to commit to the process. **This rule is hard to follow when you aren't seeing results. But it's also when the rule is most crucial.**

RULE 5: PREPARE YOURSELF

Distractions are everywhere, and let's face it – we're all prone to them. Whether it's the ping of a notification, the lure of social media or the endless to-do list, staying disciplined can feel like an uphill battle. But here's what I've learned: preparation is your secret weapon. It's the most controllable advantage you have in a world full of chaos. Preparation means setting yourself up for success by shaping your environment and priming your mindset. For example, I used to struggle with staying off my phone in the mornings. It was the first thing I'd reach for, and before I knew it, I'd wasted 20 minutes scrolling. So, I tried something simple: I placed an open book over my phone at night. In the morning, if I reached for my phone, I'd first have to read a page of the book. Nine times out of ten, that single page pulled me in, and I'd forget about my phone altogether.

Want to eat healthier? Spend an hour meal-prepping your lunches for the week. Want to visualise every morning? Stick a note on your mirror, set a phone reminder or block ten minutes on your calendar. In fact, let's put this into practice right now. Stop reading for a moment and set yourself up for success. Write a sticky note, set an alarm or prep your environment in some way. This rule has been a game-changer for me. I prepare myself to win physically, mentally and emotionally. Small acts of preparation remove friction and stack the odds in your favour. **Separate yourself from your distractions, or your distractions will separate you from your goals.**

RULE 6: START WITH ONE THING AND DO IT WELL

I didn't become disciplined overnight. It took work and intentional progress. The key is to start with something small so you can show yourself you have the skill. Make it so small that you can't say no. I would suggest applying this to something you already do in your life but want to be a bit better at. For example, 'I am going to read one page of this book every day for a week' or 'I am going to drink two litres of water every day for the next two weeks.' Then build it up with a bigger activity or for a longer period of time. Once you have shown yourself you can be disciplined, you will also build your confidence. This will be useful when trying new habits or activities like visualisation and mental fitness. When something is new, committing to it can feel even harder. **Build the foundation and competence of discipline first.**

RULE 7: REWARD YOURSELF

Discipline shouldn't be treated like a punishment – even though it can feel like it sometimes. Rewarding yourself establishes a positive reinforcement loop – your brain will create a positive association for the habit it is trying to cultivate. For me, the more I reward myself, the more disciplined I want to stay. It doesn't have to be anything massive (remember the Tiny Habits® Recipe on page 201?). Even saying a simple 'well done' internally after you complete a visualisation is enough. And even when you are finding it harder, still recognise that you are trying and honour the effort you are putting in. **Little wins, big wins – celebrate them all**.

Ask yourself, which rule do you need to work on the most? And how can you start doing that?

These rules have not only helped me build unstoppable discipline, but they have also ensured that I maintain it. However, they don't work without one particular ingredient, and that is compassion.

THE ART OF COMPASSIONATE DISCIPLINE

When it comes to building a habit like visualisation – or any habit – it's common to be our own worst critics. Every time I slipped or missed a day, I'd judge myself harshly. The cycle of guilt and frustration made it even harder to stay consistent.

Then I discovered something that changed everything: **compassionate discipline**.

At first, it sounds like a contradiction. Discipline often feels rigid and demanding, while compassion is soft and forgiving. But when you bring them together, they form a powerful partnership that makes sticking to habits not only possible but sustainable.

Let's break it down.

Compassionate discipline starts with kindness towards yourself, especially when you stumble. Too often, when we miss a day or fall short, our instinct is to criticise. But growth isn't fueled by self-punishment – it thrives on self-support. Instead of beating yourself up, practise saying, 'It's okay. Life happens. I'll start again tomorrow.'

Of course, compassion doesn't mean letting yourself off the hook entirely. If you notice that you've been off track for a week or more, it's time to reflect. Check in with your intentions, actions and execution. Ask yourself: 'Is this still aligned with what I want? Am I making it harder than it needs to be?' This reflection isn't about blame – it's about realignment.

The truth is, no one is 100 per cent disciplined all the time – not me, not you, not anyone. And that's perfectly fine. Life is about balance. Let me show you what compassionate discipline looks like in my own life:

I wake up at 6.30am for a two-hour morning routine.
But I also let myself stay in bed until noon sometimes.

I listen to two educational podcasts every day.
But I also binge-watch Netflix and *Gogglebox*.

I enjoy quiet nights in, journalling and reflecting.
But I also dance until 2am with friends.

I work long hours and weekends when needed.
But some days, I do absolutely nothing.

I wear tailored suits to meetings.
But I've also rocked up in pyjamas.

Do you see the pattern? Compassionate discipline allows me to hold myself accountable without demanding perfection. It's not about being rigid; it's about being real. And despite the flexibility, I still reach my goals, achieve excellent results and live the happiest, healthiest version of my life.

This philosophy works because it honours the human experience. Some days you're on fire, ticking off every box and crushing your goals. Other days, you're scrolling through memes or taking a nap. Both are okay, as long as you maintain a commitment to your bigger vision.

Remember, discipline isn't about sacrificing joy or spontaneity – it's about creating the life you truly want. And compassionate discipline is how you make that life sustainable.

So, the next time you falter, pause. Be kind to yourself. Realign if needed. Then pick up where you left off. Discipline and compassion aren't opposites – they're teammates. Together, they'll take you further than you ever thought possible.

TAKE OWNERSHIP

This might upset some people but: **your mental fitness is entirely your responsibility**. It's not your employer's job. It's not your family's job. It's not your coach's job. And it's not your doctor's job.

While mental weakness, limiting beliefs and negative thought patterns aren't always your fault, **they are still your responsibility**. This is a hard pill to swallow. And it took me five years to swallow it.

I used to blame everyone else for the unfair position I found myself in. It was easier that way. I blamed my parents. I blamed my doctors. I blamed God. I blamed anyone in my line of sight. I played the victim in my own life and I had every right to feel that way. But victimhood got me nowhere. Even today I still catch myself in victim mode, where I am blaming external circumstances or people for the way I feel and act. I've quickly come to the realisation that blaming others or expecting them to change a situation won't work. The only person that can really help you is you. Yes, support from others is important but ultimately, **you** are the one who has to want to get better. You have to want to take action, do the work, and become a better version of yourself.

I often see this play out in corporate environments, where employees shift blame onto the organisation:

'The company didn't train me.'

'My manager is destroying my mental health.'

'It's not my fault; the system is flawed.'

While these complaints may be valid, constantly blaming others leads to frustration and stagnation. What if, instead, you took ownership of your own growth? Even by just 1 per cent? Imagine the empowerment of learning a new skill on your own, setting boundaries or finding proactive ways to improve team communication.

There are two types of people when it comes to mental fitness. The first sees responsibility as a burden: unfair and draining. The second sees responsibility as an opportunity: fulfilling and empowering. I encourage you to be the latter. Responsibility, at its core, means having the ability to respond. Instead of giving others control, take charge of your own life.

Think about the difference between renting and owning. If you rent something – a car, an apartment, anything – you view it as temporary. It's not truly yours, so you don't care for it as much. But when you own something, it becomes part of you. You take better care of it because it's yours.

So, take ownership:

- Take ownership of your mission.
- Take ownership of your problems.
- Take ownership of your solutions.
- Take ownership of your actions.
- Take ownership of the bad.
- Take ownership of the good.
- Take ownership of your mistakes.
- Take ownership of your mind.
- Take ownership of your body.

(P.S. I can't convince you to take ownership. **You have to choose it.** But just know: if you don't, someone else will.)

BE MORE LIKE USAIN BOLT

I'm not typically one to subscribe to the idea of a single quote 'changing your life', but I can't deny that there is one that truly resonated with me. It's lived rent-free in my mind, and I want it to take up residence in yours too.

I like getting results immediately. I want to see that my inputs have yielded output. When I started practising visualisation seriously, I thought my life would change the next day. But it didn't. A week later, I still felt the same. A month later, I was ready to give up. It wasn't working. I remember being back at my visualisation teacher's house and I was so angry. I was telling her how I put in so much effort to get better; I tried so hard every day. I was being so disciplined and yet I wasn't seeing the results I wanted. She kept telling me to be patient. Later that day, I saw this quote from Usain Bolt: 'I trained four years to run nine seconds and people give up when they don't see results in two months.'

Just sit with that for a minute. In fact, read it again. Can you imagine training for four years for it to all come down to one throw, or one nine-second run, or one vault? Reading this quote gave me much-needed perspective. It taught me the power of patience.

Athletes train for years before they get the results they want. Their level of commitment and discipline is unmatched and something we can all learn from. Michael Jordan gets praised for being the best of all time, but no one talks about the fact that he played for nine years before he won any championships. We live in a time of instant gratification. Want dinner? Deliveroo in 20 minutes. Need an item? Amazon will deliver it within 24 hours. Want a date? Swipe right. This type of convenience means that when we don't get instant results, we tend to give up or think it's not working. In reality, goals and progress are not achieved overnight, and they shouldn't be. Greatness and excellence come from investing in the long game. In fact, my recovery took longer than I ever expected. It took years of deep work, resilience and small actions every single day. But because I was connected to my vision of who I wanted to be and what I wanted to do, I kept going. Along the way, my teacher taught me to stop asking, 'How long is this going to take?' and instead tell myself, 'However long it takes.' That shift changed everything for me. Some goals do get results quickly. Others don't. It's important to have both in your life.

While Bolt has taught me the power of patience, ironically he has also taught me the power of impatience. A rare trait that differentiates the good from the great is the ability to find the balance between the two. Patience is about waiting for the right moments and allowing skills and strategies to develop over time. It's about trusting the process and timing. Impatience is the dissatisfaction with slow progress. At a glance, they conflict with each other. But the sweet spot lies in being impatient with your inputs and actions, but being patient with the results and outcomes.

This is especially the case when it comes to training the mind. Think about it. You are rewiring this incredible asset you have that has been functioning for the last 20, 30, 40 or more years on a certain program. That's not going to all change tomorrow. Most of us are impatient about the results and patient about our inputs. It's time to flip that mentality. Become patient for your long-term goals, where you can maintain composure and enthusiasm while you overcome setbacks and challenges. Be impatient in your efforts, your everyday actions and your hunger to improve. When you do this, you strike a subtle duality that creates an incredible momentum.

So, as you embark on your mental fitness journey, just remember to be more like Usain Bolt – maybe he can be your Batman Effect character (see page 117).

PROGRESS IS GREATER THAN PERFECTION

There is no such thing as a perfect visualisation or perfect mental fitness. That would be the same as saying there is a perfect walk or the perfect run. Of course, there are better techniques and forms one can have, but there is no such thing as doing it perfectly. When it comes to training your mind and integrating mental fitness into your life, you want to shift your mindset from perfection to progress.

I am often asked what progress looks like. But, as we can't see the mind, it can be difficult to know if something is working or what we should be looking out for. Progress will also look different for everyone. This is because your goals are different, your mind is different, who you are is different and most likely how much time and money you have to invest will also be different.

That being said, whether you are just starting your visualisation journey or you are mastering your practice, there are many different indicators of progress. This includes progress mentally, physically and emotionally. Generally, in two weeks you will start to notice changes in the way you think and feel; in about four weeks you start to see changes in your performance, mindset and sustained emotional management; and then from 8 to 12 weeks onwards, your new way of thinking, feeling and performing becomes like second nature. Studies have indicated that you need at least 90 days to rewire the brain and achieve measurable change. Of course, this is going to depend on a lot of other factors, and some complex changes will require longer, but generally I find this to be an encouraging amount of time to commit to.

> Below, I have listed the most common changes people see once they start practising visualisation, based on research as well as my coaching, events and personal experience.
>
> ### Think
> — better clarity in thinking and goal-setting

- improved cognitive ability to make decisions (faster and sharper)
- better awareness and intentionality around self-talk (kinder, more motivating)
- being less reactive to triggers and instead able to respond more intentionally
- approaching challenges with optimism rather than fear
- more visionary thinking about goals (individual and company level)
- enhanced creativity for projects and ideas
- better awareness and implementation for taking pauses and rest
- looking for the good before the bad and being able to reframe situations better

Feel
- better ability to relax and calm the body reactively and proactively
- noticing your emotions don't consume you as much; ability to feel and release them
- other people's opinions of you don't change how you feel about yourself
- reduction in anxious feelings and having a better relationship with them
- feeling more connected to the body
- feeling more secure and confident in oneself
- more regulated nervous system (out of fight, flight or freeze mode)

- noticing and feeling gratitude more often
- higher motivation and hope for the future
- increased drive and determination towards goals
- less physical pain, or ability to manage it better
- considerable change to the energy you bring to the world

Perform
- ability to handle pressure better and maintain good performance
- an increase in standard of delivery or performance (sport, skills, tasks)
- experiencing fewer nerves or less anxiety before presentations/matches
- better ability to reflect after and implement improvements
- speaking up more and standing up for oneself
- implementing more boundaries without feeling guilty
- clear improvement in sports and fitness skills (faster, stronger, better technique)
- ability to focus better on tasks, a change in productivity, reaching and sustaining flow state
- experiencing less burnout and sudden breakdowns

Everyone is different, but the key is to become curious about the way you think, feel and perform. But overall, look for steady, gradual progress.

Remember, though, that progress is never linear, especially with the mind. Even when you don't see obvious signs of improvement, you are still laying the foundations for future progress and results. I think that's one of the hardest aspects of training the mind – you can't always see what's happening. That's why it's even more important to spend time reflecting and understanding yourself. The changes can be so subtle that we can even miss them. In fact, I was recently on a call with one of my coaching clients who I have been working with for over a year. During the session, she was explaining a recent situation she was dealing with at work. There was a lot of conflict and stress. As she told me how she handled it, I pointed out how different her response was from a few months ago. She paused and said, 'I didn't even notice that.'

Often, we are so caught up in life that we forget to pause and think about how far we have come. I am a massive culprit of this. Even after going through difficult times with the illness, I still get caught up in the drama of life. When I zoom out and realise the progress I have been making, I am able to ground myself and appreciate the work I am doing. A practical action I would suggest doing is periodically asking yourself 'Where was I then versus now?' or 'How would I have reacted to this three months ago?' This will encourage you to notice how far you have come.

A FEW FINAL WORDS

*If you're going to question anything, question
your limits, not your potential.*

Congratulations! Reaching this point means you've taken a huge step forward. Let's take a moment to acknowledge that. You invested your time, energy and focus to explore the tools and techniques that can transform your mental fitness. Many people talk about changing their lives, but few take action. You did.

This is your moment to reflect on what you've learned and, more importantly, what you've accomplished. Every exercise you've tried, every thought you've challenged and every question you've explored has planted a seed. Let these new ideas and ambitions take root. Change doesn't happen overnight, but by showing up for yourself and practising the visualisations, you've already started to build the strongest version of you.

It's easy to look at champions – whether they're elite athletes, world-class performers or even the high achievers you see around you – and assume they have something special that you don't. But I'll be the first to tell you that they're not born with some mystical power or superhuman talent. What sets them apart is skill. Specifically, their mental skill, and the determination to train it.

Confidence? It's a skill. Staying calm under pressure? That's a skill. Being resilient, kind, creative or focused? All skills. Just like learning to ride a bike or play the piano, these traits are trainable. One of my clients was reluctant to start because he believed he was too far gone. That it wasn't possible to reshape his thinking after the decades of negative patterns and self-doubt. But the human mind – your mind – is incredibly adaptive. It can rewire itself with the right training and persistence.

Everyone who reads this book will experience it differently. For some of you, coming to the end may feel a bit overwhelming – like there's so much to work on that it seems hard to know where to start (the 30-day challenge on page 232 will help with this).

But you also might have felt a spark. Maybe it was a flash of excitement, a stronger determination to go for your goals, a new belief in yourself or even a sense of relief after visualising. As you move forward, remember that mental fitness is a lifelong process. Even though I'm considered an expert in this area, I'm still learning and building my own mental strength every day. There's no 'finish line' to this work. It's a constant process of refining, trying and experiencing. Sometimes, you'll feel progress; other times, you'll feel like you're standing still. That's part of growth. The important thing is to stay consistent and keep challenging yourself to build a mindset that supports your highest aspirations. I strongly encourage you to return to this book as a resource. Revisit the exercises, re-read the parts that resonated most with you and keep visualising the life you want to lead.

We live in a world that tempts us with quick fixes: a pill, a weekend retreat, a new brain-hacking device. We chase the latest and shiniest trends, from AI to psychedelics, expecting them to change us overnight. But what about the basics? The basics work. They always have and they always will, and that's why this book has focused on them.

While this book draws on stories of athletes and champions, remember that visualisation is not just for the next Phelps or Ronaldo. It's about becoming the best version of *you*, 0.1 per cent at a time. Everything you need to become mentally fit is between your two ears. You have the most powerful asset in the entire world and yet the average person would rather train ChatGPT than their own mind. So, instead of visualisation being this technique that only elite athletes and CEOs use, it's time you take full advantage of it and become the champion you are. One of the biggest issues we face in the world today? How much we focus on other people's lives compared to our own. Start from where you are and with what you have. You can use your mind to worry, stress and overthink or you can use it to believe, create and focus. The choice is yours.

To close, I want to return to this thought I shared at the start:

VISUALISE

Your mind is like a puppy. If you don't train it, it will sh*t everywhere. But if you take the time to guide it, nurture it and teach it, it will become your greatest companion — loyal, powerful and capable of helping you achieve anything.

Thank you for trusting me to guide you on this part of your journey. I can't wait to see who you become.

APPENDIX

WELCOME TO YOUR 30-DAY VISUALISATION CHALLENGE

When it came to designing this challenge, I wanted to make it as tangible and measurable as possible. In 30 days, I want you to get from A to B. It's an opportunity for you to try a variety of visualisations in a structured way, while also directing them towards a goal or purpose. You can pick any goal. It can be to become more confident or it could be to improve your public speaking and presenting skills. Maybe it applies to something more specific, like an interview you have coming up or a marathon you plan to run at the end of the month. If you want to, you can even pick a longer-term

WELCOME TO YOUR 30-DAY VISUALISATION CHALLENGE

goal like starting your own company. It might be winning Wimbledon. It could be overcoming an illness and feeling healthy again.

If you choose a short-term, tangible goal, at the end of or even during the 30-day challenge, you are likely to achieve it. If it is a longer-term goal, you will make key progress towards it, but may not complete it. Either way, try to pick one thing to focus on.

I want to make it very easy for you to get started. All it requires is a minimum of a few minutes a day. But you have to commit to the 30 days. This won't work if you do it sporadically. What you put in, you will get out.

We underestimate how much we can change our mind, our attitude and our character in a month. The more you train, the more you visualise and the more you reflect on your progress, the more change you will see. As you proceed with the challenge, refer back to Part 2 for the visualisations. And if you want to accompany the challenge with daily emails and visualisations, head to mayaraichoora.com/visualise-extras.

Ready? Let's begin ...

Day 1: Set your goal

What is the goal or project you are working on? The thing that excites you? Pick something slightly challenging yet possible. Write down the specific goal you want to achieve in the next 30 days. Next, write down why and, finally, how you want to feel doing it. Choose a time you will visualise each day.

Day 2: Outcome visualisation for inspiration (4–6 minutes, see page 110)

Visualise yourself achieving your goal. See the results in advance. Imagine the details, including the environment, the people, the activities and what the end result looks like. Feel the motivation and inspiration of the goal.

Day 3: Outcome visualisation with more clarity (8–10 minutes, see page 110)

Refine your vision of the outcome and improve the visualisation. Focus on adding more sensory details like what you can hear, taste and smell. Immerse yourself in the goal. Make it so real for your brain. Remember, mental imagery is a skill. Practise making the vision clearer and clearer.

Day 4: Outcome visualisation with emotional connection (8 minutes, see page 110)

Build on the previous days by adding more emotion to the visualisation. Tune into how you want to feel once you have achieved the goal. Where do you feel it in your body? Expand it, make it bigger. Allow yourself to smile, laugh or cry. Be open.

Day 5: Visualise your most confident self (5 minutes, see page 116)

Who do you need to be to achieve your goal? What part of your character do you need to level up? Your confidence?

Kindness? Discipline? Visualise this version of you in advance. What do they look like and how do they act? Bring this into the present moment and commit to this version of you for the rest of the day.

Day 6: Batman Effect
(7 minutes, see page 119)

Let's go one step further. Today, I want you to choose your character. It can be fictional, someone who inspires you or even an animal. Get creative. Make a list of your character's attributes. Visualise this version of you. Really push yourself here: be the fiercest, most confident, powerful, peaceful and brave version of you. Try to apply your character to a particular task today and notice the difference. And remember, during the day ask yourself, 'What would X do?'

Day 7: Outcome visualisation and reflection
(10–12 minutes, see page 110)

Combine your goal visualisation with your character. First, mentally rehearse your goal, then who you need to become and then go back to your goal. See if you can push yourself in this one. What would 1 per cent bolder look like? Or 1 per cent more confident? Or 1 per cent more relaxed?

Take a moment to reflect on the last week. What have you noticed about how you have felt? Did you act differently in some situations? What about your motivation and confidence? Do you want to change anything in your character?

Also take this time to reflect on the routine you are building. Is the time working? Are you showing compassion? Are you keeping the promises you make to yourself?

Day 8: Process visualisation
(8 minutes, see page 132)

Now it's time to shift the focus from the outcome of the goal to the steps required to achieve it. Visualise the day ahead. Focus on the daily actions you need to do to reach your goal. See yourself performing these actions in the best way you can. As well as work-related actions, this also includes social and personal ones.

Day 9: Level up your performance
(6–8 minutes, see page 132)

Let's apply process visualisation to a specific task or activity. Pick one thing you are working on in the next few days – something you want to excel in and improve your performance in. For example, an interview, a presentation, a golf game, a training session, hosting a dinner and so on. Mentally rehearse the action again and again. See yourself doing it exactly the way you want to. And remember, the mind is the limit. So push yourself a little. Add emotion and details. Feel the movements, pace and interactions too.

Day 10: Combine outcome + process visualisation
(1–5 minutes, see pages 110 and 132)

Start by repeating your outcome visualisation for your goal. See the results in advance and feel the emotional connection to it. Now, add a process visualisation on top. See the actions you are taking in your day, focusing on the performance. Add emotion and set your intention.

Day 11: Focus and productivity
(4 minutes, see page 143)

Today, I want you to practise the skill of being focused. What do you need to complete today? What are some of the tasks you have been avoiding? Pick one or two activities or tasks and mentally rehearse them with a high level of focus. The aim is to see yourself being the focused version of you and getting it done. Take this into your day and watch how much more you complete.

Day 12: Letting go
(5 minutes, see page 152)

We hold so much in our mind and body. Use today's visualisation to help you let go of anything you don't need. This can be mental, emotional, physical or even a person/situation. Follow the process of imagining writing it on paper and placing it in a river. You can mentally repeat 'I let go' to help the mind process it.

Day 13: Combine outcome + process visualisation
(15–18 minutes, see pages 110 and 132)

I want you to repeat the visualisation from Day 10. Generally, outcome and process is a good non-negotiable visualisation to do in the morning. Add more detail and emotion and make the processes even more effortless for the mind. Get into the habit of starting your day like this.

Day 14: Reflection and rest
(10 minutes)

How was your experience with process visualisation? Write down any changes in your productivity, focus and performance. Also take this time to rest the mind.

Day 15: Creative visualisation
(3–6 minutes, see page 159)

It's time to bring in some creative visualisation. When we are progressing towards a goal, it is normal for different emotions and feelings to come up. Use today's practice to locate an uncomfortable/hard feeling that you have either been avoiding or that's at the surface. Locate it in your body, add the shape/colour/character and slowly let it dissolve. Stay really aware of how you feel before and after.

Day 16: Relaxation visualisation
(5–8 minutes, see page 166)

We can always use more opportunities to de-stress and relax. Today is about practising moving from your sympathetic nervous system to your parasympathetic nervous system. Whether you are feeling stressed or not, this will calm the mind and body. Visualise yourself at a very peaceful lake or river. If there is another environment that suits you more, feel free to pick that. As mentioned, do what works for you and make it personal!

Day 17: Outcome + gratitude visualisation (12 minutes, see pages 110 and 161)

Today, we are combining your goal visualisation with a creative visualisation for gratitude. Start by connecting to your goal (hopefully this one is getting clearer and clearer for you). After seeing the end result, I want you to take a few minutes to feel gratitude in your mind and body. Create a shape/colour/character and expand it in your body.

Day 18: Outcome + process + creative gratitude visualisation (15–20 minutes, see pages 110, 132 and 161)

This is the longest visualisation so far. Repeat your goal visualisation, focusing on the results. Then add in the processes of your day or particular tasks. End with a creative visualisation (page 161), expanding and feeling a deep sense of gratitude for yourself and what's to come. If your mind wanders, bring it back to a clear image. Don't worry if you can't do the full 15 minutes. Do as much as you can.

Day 19: Express creative visualisation (1 minute, see page 159)

Creative visualisation is extremely effective in 40–90-second bursts. Today, I want you to practise mini ones for your emotions. This in itself is a skill. You can either pick the same emotion, like anxiety, and rehearse it again and again, or you can pick different emotions to see how they feel different. You can keep doing this throughout the day as many times as you like.

Day 20: Relaxation + goal visualisation (8–10 minutes, see pages 166 and 110)

Today is another combination of creative and outcome. Start with a relaxing lake visualisation and follow it with your vision and goal. See if doing extra relaxation beforehand helps you concentrate or connect more with the goal.

Day 21: Reflection and rest (5–8 minutes)

How was your experience with creative visualisation yesterday? How did you manage your emotions and stress? Feel free to note down some of the colours, shapes or characters you saw. Which one did you repeat the most? Do you need to do more work there? What changes have you noticed? Take a day to rest the mind.

Day 22: Negative visualisation (6–8 minutes, see page 174)

Naturally, there might be some challenges or worries that arise in regards to reaching your goal. Today, I want you to try a negative visualisation and see yourself overcoming the challenge. It can be tiny or big. Or perhaps there is a habit you are trying to break. Remember, the key is seeing yourself overcoming the obstacle.

Day 23: Negative visualisation for goals
(8 minutes, see page 174)

In this visualisation, I want you to see what would happen if you didn't achieve your goal. How would it make you feel? What is the risk of it not happening? This often makes people more motivated and ready to take action. So, for example, if your goal is to lose weight, visualise yourself gaining weight and not achieving the goal. Does it make you want to make it happen even more? Remember, this might not be for everyone! Stay curious.

Day 24: Explorative visualisation
(6 minutes, see page 188)

Pick any creative task you need to do today. It could be designing a slide, writing content or even coming up with an idea for dinner. The aim of today is to practise the technique and push your mind to be more creative with its ideas. If you don't have a creative task to do, use your goals as one. Explore some new goals you might want to try. Remember, your imagination has no limits! At the end, take some notes of the things that came up for you.

Day 25: Mentally rehearse your day
(4–6 minutes, see page 145)

Repeat a process visualisation for your day. Add in the activities you want to complete.

Day 26: Build your own! (6–10 minutes)

You've been trying out different techniques over the last few weeks. Today, I want you to take ownership of the routine you want to try. Feel free to combine different ones or keep it simple and use one type.

Day 27: Outcome + process visualisation (15 minutes, see pages 110 and 132)

Practise your non-negotiable visualisation. Keep building the beliefs. If you want to, add some affirmations or experiment with music.

Day 28: Confidence/Batman Effect + gratitude visualisation (10 minutes, see pages 116, 119 and 161)

We are nearly at the end of the 30-day challenge. I want you to take some time to connect with your character again. Notice how familiar they may feel. Continue to watch how they show up and spend a few moments feeling gratitude for them.

Day 29: Outcome + process + gratitude visualisation (15 minutes, see pages 110, 132 and 161)

Practise your non-negotiable visualisation.

Day 30: WOOHOO! You did it!

Congratulations! You have completed 30 days of visualisation. Take a moment to reflect on how visualisation has impacted your mindset and actions. Think about

WELCOME TO YOUR 30-DAY VISUALISATION CHALLENGE

the challenges you have overcome, the success you have achieved and the parts of your character you have been proud of. Also think about what you want to spend more time on improving or elements that you found tough. It's okay if it doesn't click right away! Celebrate your achievements and commit to continuing your visualisation practice beyond the challenge.

Now it's all about the next 30 days. This is where people tend to lose momentum and discipline. To stay focused and committed to your goal, put reminders in your calendar or set alarms at your chosen time. Choose your non-negotiable visualisations and continue to build from there.

RESOURCES

David Hamilton, *Your Mind Can Heal Your Body* (Hay House, 2008)

Norman Doidge, *The Brain That Changes Itself: Stories of Personal Triumph from the Frontiers of Brain Science* (Viking, 2007)

Tara Swart, *The Source: The Secrets of the Universe, the Science of the Brain* (Ebury, 2019)

To join the 30-day challenge with daily emails and visualisations, follow this link: mayaraichoora.com/visualise-extras

ACKNOWLEDGEMENTS

There are so many people I want to thank who have been part of my journey, but that would be a whole other book so here are a few I want to mention:

To my mum and dad, who stood by me at my worst and never stopped believing in me. You've shown me what true unconditional support and love looks like, and I'm forever grateful for you both.

To my sisters, Rakhee and Sona, for being strong women I look up to and draw constant inspiration from. Your strength, resilience and selflessness inspire me every day.

To my agent Tom and the team at Penguin Random House, including the editors, designers and everyone who poured so much care and attention into this book. Your dedication has brought this project to life in ways I never could have imagined.

To Hannah, for the beautiful illustrations that brought this book to life. Your creativity and care have added so much depth to this project.

To Pravina and Yogini, my coaches, whose wisdom has shaped me and this work in ways that continue to transform my life. I wouldn't be here without either of you.

To Jaznique, for making me feel like I am capable of anything. You remind me how much magic there is in this world.

To Ben, for your unwavering kindness, support and dedication.

To Kesta, for making life feel like the best adventure ever.

To Isi, for always making me feel loved and seen.

To Kate, for growing with me through every chapter.

To Hanna, for being a guiding light in my life.

To Liv, for your deep empathy and compassion.

To Emma, for accepting me in my brightest and darkest moments.

To Ethan, for reminding me to be brave.

To Siddhi, for always encouraging me to dream bigger.

To my doctors, health professionals and carers along the way, thank you for always doing your best.

To the nurse who asked me the question in hospital, I will never forget the power of a question. Thank you.

And finally, to every reader who's picked up this book, thank you for allowing me to share my story with you. Your time, trust and openness are the greatest gift of all.